Learn Visio 5.0

For users of
Visio Technical and
Visio Professional

by Ralph Grabowski

Wordware Publishing, Inc.

Library of Congress Cataloging-in-Publication Data

Grabowski, Ralph
    Learn Visio 5.0 / by Ralph Grabowski.
      p. cm.
    Includes index.
    ISBN 1-55622-568-7 (pbk.)
    1. Mechanical drawing. 2. Visio.
    I. T353.G68   1997
    604.2'0285'5369--dc21

        97-43820
        CIP

ISBN1-55622-568-7
10 9 8 7 6 5 4 3 2
9711

All inquiries for volume purchases of this book should be addressed to:
Wordware Publishing, Inc.
2320 Los Rios Boulevard, #200
Plano, Texas 75074
Telephone inquiries may be made by calling: (972) 423-0090

Contents

Contents

Introduction

Getting Started

Visio is a technical drawing program. Unlike other technical drawing programs, such as computer-aided drawing or drafting software, Visio is easy to use. To make a drawing, you don't even have to know how to draw! You simply drag shapes from stencils onto the page. When you prefer to draw freehand, Visio includes a number of tools for drawing lines, circles, boxes, and curves.

Visio creates intelligent drawings. The shapes are called *SmartShapes* because they know where they connect with each other. All shapes have a minispreadsheet hidden behind them, called a *ShapeSheet*. This spreadsheet contains all known information about the shape and can be manipulated directly. Lines are known as *SmartConnectors* because they stretch when you move any connected shape.

Visio is available for three operating systems: Windows v3.1, Windows 95, and Windows NT. Visio operates identically under each; the primary difference is that the Windows v3.1 version is limited to Visio 4.0 and will not be upgraded. Visio 5.0 is the current version for Windows 95 and NT.

Which Visio is For You?

Visio 5.0 is available in three flavors: Standard, Technical, and Professional. Which product is best for you? The following table shows the primary differences between the three versions of Visio 5.0. Note that Technical and Professional contain everything found in Standard.

Visio v5.0	Standard	Technical	Professional
Users	*General business users:* Administration Finance Human resources Sales and marketing	*Engineering:* Drafters Facilities managers Technical documentation specialists	*Information technology managers:* Database consultants Network administrators Software designers
Purpose	Create flowcharts and business diagrams.	Create 2D technical drawings and schematics.	Document a corporation's information systems and processes.
Shapes	Basic networks Block diagrams Flow charts Organization charts Project timelines Geographic maps Marketing shapes Office layouts TQM diagrams	*Same as Standard, plus:* Electrical and electronic Facilities management Buildings controls-automation Fluid power Home and landscaping HVAC Manufacturing & assembly Mechancial engineering Process plant Space planning	*Same as Standard, plus:* Database design Internet design Network design and documentation Software design and visual modeling
Wizards	Database connectivity Flowchart Organizational chart Page layout Project timeline Property reporter SmartShape	*Same as Standard, plus:*	*Same as Standard, plus:* Database creation Flowchart database Map database Network database Network diagramming Web diagrams
Add-Ons		AutoCAD file converter Area analysis Netlist generator Valve builder	. . .

This book covers all three versions of Visio 5.0: Standard, Technical, and Professional. Most of the time, this book discusses Visio features that are common to all three programs. The notes Visio Technical Only and Visio Professional Only indicate the few times when the discussion is specific to Visio Technical or Visio Professional.

Starting Visio

To start Visio under Windows, either (1) select **Start | Programs | Visio 5.0**; or (2) double-click the shortcut icon on the Desktop. (To exit Visio at any time, press **Alt+F4** or select **File | Exit**.)

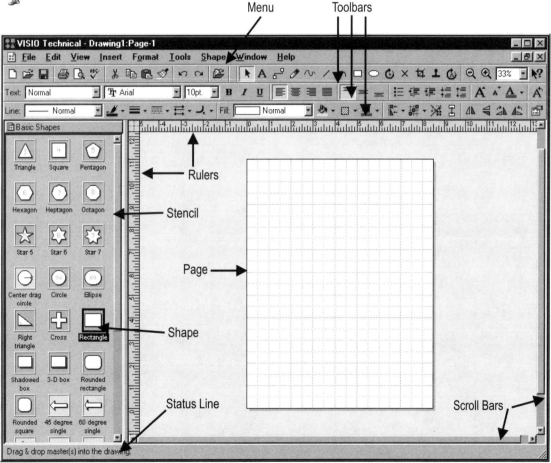

A Guided Tour

When Visio first starts, there isn't much to see. Select **File | New | Browse Sample Drawings** to open one of the sample drawings. Once a drawing is open, the Visio screen comes to life, as shown in the illustration on the preceding page.

Along the top is the menu bar. The ten menu items contain all of Visio's commands. Below the menu bar are the *toolbars*. You have a choice of displaying seven toolbars. By default, three toolbars are displayed: Standard, Text, and Shape. To add and remove a toolbar, right-click on the toolbar and select (or deselect) the name from the popup menu: Standard, Text, Shape, View, Web, Page, and Developer.

At the left is a *stencil* with *shapes*. You can open more than one stencil at a time, or you can have no stencil open. Each stencil holds 25 to 35 shapes. To use a

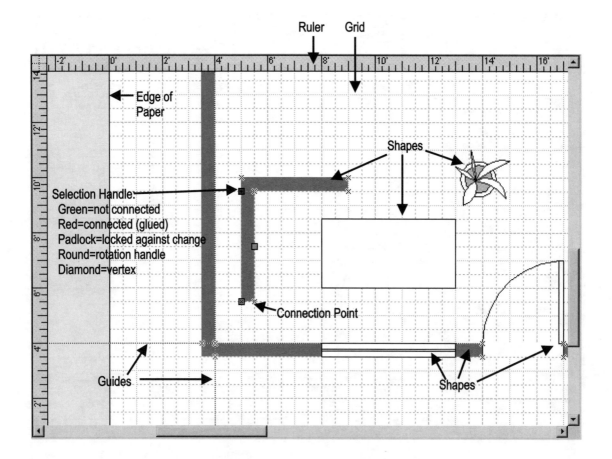

shape, drag it from the stencil to the page. *Scroll bars* let you move the page around the window. The *rulers* measure distances.

The largest part of the Visio window is the drawing area, which displays a page. While most of the time you will probably work with a standard 8½ x 11-inch page, Visio accommodates a wide variety of page sizes—all the way up to 36 x 48 inches.You can have up to 200 pages in a drawing, although only one page is displayed at a time.

Visio uses many visual hints to help you draw quickly and accurately:

Grid: On a blank drawing, all you see initially are the grid lines. The grid helps you position objects; you can choose to have shapes snap to the grid for accurate horizontal and vertical positioning. You may change the grid spacing, including independent x- and y-spacing.

Ruler: The ruler reports the size of the drawing in scaled units. By holding down the Ctrl key, you can drag the zero-point (from the intersection of the two rulers) to a new zero point. Like the grid, the ruler helps you position objects; you can have shapes snap to the ruler increments for accurate horizontal and vertical positioning.

Connection point: Small blue x shapes appear whenever you select a shape or object. These indicate the points where you can connect one object to another.

Selection handle: Perhaps the most important visual clue is the selection handle. Its shape and color change, alerting you to its properties. The most common selection handle is a small, green square. This indicates the points where you can change the size and rotation of the object. Click and hold the square, drag (to change the size or the angle), then let go of the mouse button. In brief, selection handles can take on the following colors and shapes:

➤ Green square: selection handle.

➤ Red square: glued connection.

➤ Padlock: shape cannot be resized.

➤ Round: a handle for rotating the object.

➤ Diamond: a vertex point on object such as splined curves.

The red color indicates the object is glued to another object. When two objects are glued, moving one object moves (or stretches) the other objects.

Drawing with Visio is Better and Faster

 You can create a drawing with Visio in exactly the same manner as with other drawing and CAD software: draw lines, circles, and other shapes, then edit them. In addition, Visio offers a much more powerful—and faster—method of drawing rarely found in other software. Follow these easy steps:

Step 1: Start a new drawing.

Step 2: Open the appropriate stencil(s).

Step 3: Drag shapes into the drawing.

Step 4: Connect shapes as required.

Using the predrawn shapes saves you the time it takes to draw objects from scratch. If the shape isn't exactly what you need, any shape is easily edited.

Connections between shapes work like magic. Small green squares tell you where the connections are located on each shape. When two shapes connect, the square turns red. When you move one shape, the other shape moves with it—until you decide to disconnect them. Visio includes many tools for accurate, automatic positioning and shape manipulation.

When you have drawings that were created in other software programs, you can import them into a Visio drawing. When you have data that resides in database files, you can link the data with Visio drawings, then access the database from within Visio. Visio can export its drawings in many vector (CAD) and raster (paint) formats; Visio can also export its drawing as a Web page.

Visio is a powerful program for working with graphics and data. This book shows you how easy it is to harness that power.

A Brief History of Visio

ShapeWare Corp., as Visio Corp. was first called, was founded in 1990 by two of the founders of Aldus Corp. (of PageMaker fame; later merged with Adobe, of PostScript fame). When Visio v1.0 was introduced in 1992, the software quickly became popular because it did not present a blank page to the new user. Instead, it invited the user to drag shapes and drop them into the page.

In 1993 ShapeWare began shipping optional stencils and shapes to Visio called "Visio Shapes." When it was renamed the Visio Solutions Library in 1996, the library included add-ons developed by ShapeWare and third-party vendors. One example is the Visio Business Modeler, which lets you analyze business models found in the SAP R/3 Reference Model.

After shipping Visio v1, v2, v3, and v4, ShapeWare began creating specific releases of Visio. Visio Technical was introduced in 1994 as a companion software to CAD products (the first version of Technical was called "version 4.1"). In 1995, ShapeWare Corp. changed its name to Visio Corp. and went public on the NASDAQ stock exchange under the symbol "VSIO."

Visio Professional was released in 1996 for IT professionals (the first version of Professional was called "version 4.5"). In 1997 Visio delivered Visio Map for GIS users, with mapping technology licensed from ESRI, one of the largest GIS software companies.

In 1998 Visio was due to begin shipping IntelliCAD (aka "Phoenix"), an AutoCAD-compatible CAD system Visio obtained by purchasing Boomerang Technology, a former division of Softdesk, which had merged with Autodesk a year earlier.

The Visio Business Partner program helps out third-party developers, many of whom customize Visio for their in-house requirements. For more information, check the Visio Web site at **http://www.visio.com**.

About the Author

Ralph Grabowski is the author of 32 books about the Internet, computer-aided design, and technical drawing, including the Learn AutoCAD in a Day series for Wordware Publishing. Ralph is the former Senior Editor of *CADalyst Magazine* and now heads up XYZ Publishing, Ltd. He is a frequent contributor to *Cadence*, *InfoWorld*, and *AutoCAD User* magazines. Ralph is the editor of *upFront.eZine*, a weekly newsletter distributed by email. He can be reached via email at **ralphg@xyzpress.com**.

Module 1

Starting a New Drawing

File | New

Uses

Before you can start drawing, you get a sheet of paper; in the same way, you start with a new drawing page in Visio. The **New** selection of the **File** menu creates a new Visio drawing. A new drawing is a blank sheet of paper. Visio 5.0 lets you start a new drawing in three different ways:

➤ Begin with a template drawing.

➤ Start with a blank drawing.

➤ Create a new drawing with a wizard.

Starting with a Template Drawing

In your office, you probably work with many kinds and sizes of paper: plain paper, memorandums, graph paper, legal size, broadsheet, forms, and so on. In the same way, Visio provides many kinds of "paper" for creating new drawings. These are called *drawing templates*. When you start Visio for the first time, it displays the **Choose a Drawing Template** dialog box. An illustration of this dialog box is shown at the top of the next page.

1

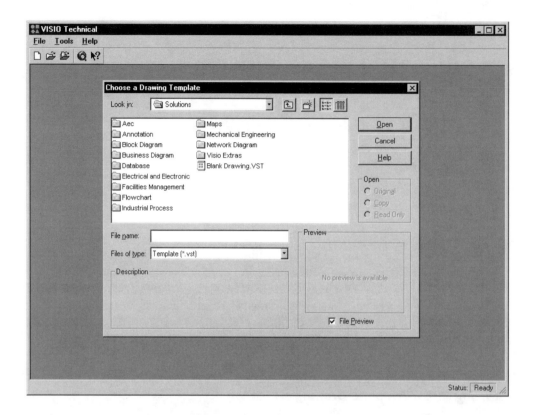

A drawing template has some drawing properties set up for you, such as the size and orientation of the paper, a scale factor, and one or more stencil collections of shapes. A template drawing has the filename extension of VST, such as **SitePlan.Vst**.

Depending on the edition of Visio you have, Standard, Technical, or Professional, each comes with a different collection of template files. Visio Standard includes templates for creating drawings of block diagrams, flowcharts, maps, office layouts, and project timelines. Visio Technical adds technically oriented templates, such as chemical and petroleum engineering, house and landscape planning, software and network design, mechanical and electrical engineering, and space planning. Visio Professional has all the templates in Standard, plus additional templates for networks, Internet design, database design, and visual modeling (software design).

For example, the **Home-Bath and Kitchen Plan.Vst** drawing template (look for it in subdirectory \Visio\Solutions\Aec with Visio Technical) automatically sets up these drawing properties:

Paper: 8-1/2" x 11", landscape orientation.

Scale: 1/4" = 1' (notice the foot markings on the two rulers).

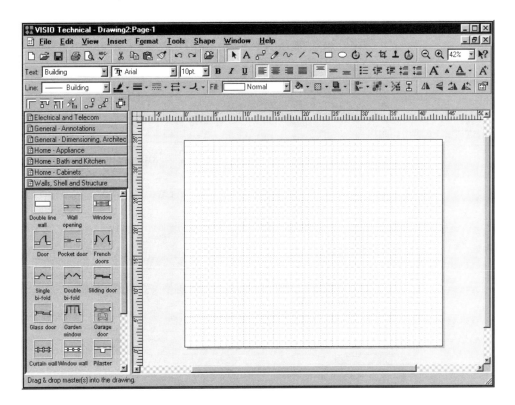

Stencils: Walls, Shell and Structure; Home - Cabinets; Home - Bath and Kitchen; Home - Appliance; General - Dimensions, Architectural; General - Annotations; and Electrical and Telecom.

Additional Toolbar: Wall Tools (located just above the stencils).

You are not stuck with the drawing templates that Visio provides. Feel free to modify any template to suit your need; just remember to give it a different name or store it in another subdirectory. Or, you can create template drawings from scratch. (You find the details of how to save a template drawing in Module 2, Saving Files.)

By creating your own templates, you create a consistent look for all your drawings. Templates are an excellent way to affirm corporate standards. For example, to ensure the corporate logo and copyright statement appear in every drawing,

3

create a layer with that information, then save the drawing as a template. The properties you can save in a template drawing include:

➤ Layers, scale, and page settings (see Module 4, Setting Up Pages).

➤ Snap and glue settings (see Module 6, Rulers, Grids, and Guide Lines).

➤ Color palette (see Module 13, Formatting Shapes).

➤ Shape styles and text styles (see Module 15, Creating Styles).

➤ Print setup (see Module 20, Printing Drawings).

➤ Window size and position.

Starting with a Blank Drawing

When you choose no template drawing, Visio starts with a blank drawing: an 8-1/2"x11" sheet of paper in portrait orientation, a scale of 1:1, and no stencil or shapes loaded. (This is unlike previous versions of Visio, which loaded the **Basic Shapes** stencil.) Starting with a blank drawing is best when you want to create a new drawing from scratch, with no assistance from Visio's templates or wizards.

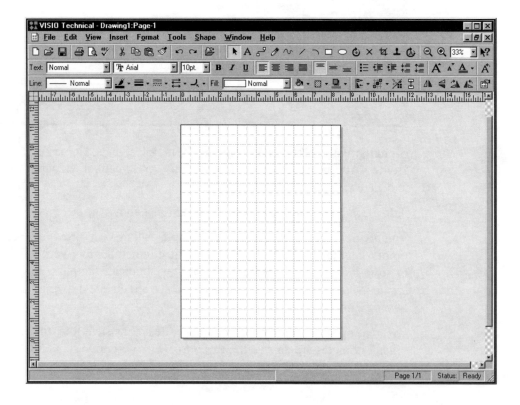

Starting with a Wizard

All Visio packages include wizards that guide you through the steps of setting up a preliminary drawing, setting the scale, filling in the title block, and placing a border around the drawing. Along the way, you are prompted to fill in information and select options. Wizards are available for creating flowcharts, office layouts, organizational charts, project timelines, and other specialized drawings.

There are two disadvantages to using a wizard: (1) you may find it becomes tedious answering the wizard's many questions, then waiting for the time it takes the wizard to complete its work; and (2) you sometimes end up doing more work editing the drawing created by the wizard than you would have starting the drawing from scratch.

Procedures

When you start Visio, it automatically displays the **Choose a Drawing Template** dialog box.

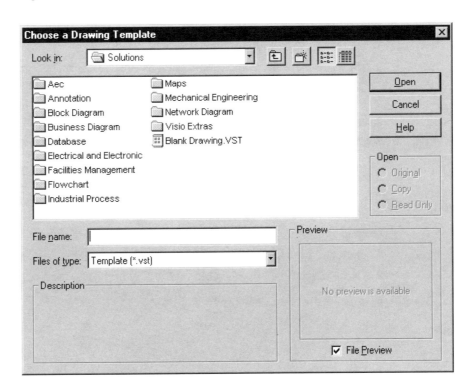

From this dialog box, you select the name of the template. Alternatively, you are allowed to start a new drawing anytime you are in Visio with the New command. The shortcut key is:

Function	Keys	Menu	Toolbar Icon
New	Ctrl + N	File | New	▣

Starting a New Drawing with a Template

Use the following **procedure** to start a new drawing with one of Visio's templates:

1. Select **New** from the menu bar.
2. Select **Browse Templates** to display the **New Drawing** dialog box, which looks identical to the **Choose a Drawing Template** dialog box.
3. Select a VST template file from the list under **Look In**. (It may be necessary to double-click a folder to find a specific template.)
4. Click the **OK** button.
5. Notice that Visio opens a blank, scaled drawing with the appropriate stencils.

Alternative method:

1. Select **New** from the menu bar.
2. Select a name from the menu, such as AEC, Block Diagram, or Business Diagram. (The list of names varies, depending whether you have Visio Standard, Technical, or Professional.)
3. Select a template name from the list. For example, under Business Diagram, you can choose from Charts and Graphics, Form Design, and Marketing Charts and Diagrams.
4. Notice that Visio opens a blank, scaled drawing with the appropriate stencils.

Starting a New Drawing Without a Template

Use the following procedure to start a new drawing without a template:

1. Select **New | Drawing** from the menu bar.
2. Notice that Visio opens a blank drawing and no stencil page.

Starting a New Drawing with a Wizard

Use the following procedure to start a new drawing with a template wizard:

1. Select **New** from the menu bar.
2. Select a name from the menu, such as AEC, Block Diagram, or Business Diagram. (The list of names varies, depending whether you have Visio Standard, Technical, or Professional.)
3. Select a wizard name from the list. For example, under Business Diagram, you can choose from Office Layout Wizard, Project Timeline Wizard, and Organization Chart Wizard.
4. Notice that Visio opens a scaled, blank drawing with the appropriate stencils, and starts the wizard.
5. Follow the instructions provided by the wizard's dialog boxes.

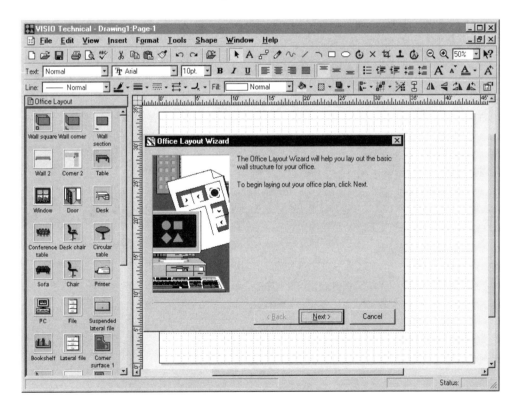

Hands-On Activity

In this activity, you use the New function to start a new drawing.

1. Start Visio to begin the activity.
2. Notice that the Choose a Drawing Template dialog box is displayed.
3. Double-click the **Business Diagram** folder to open it.
4. Double-click **Charts and Graphs.VST** to select it.

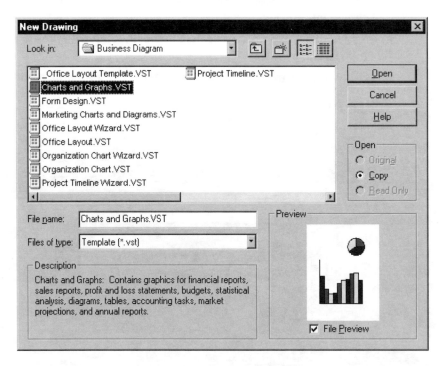

5. Click **OK**. Visio opens a new drawing that looks like a sheet of graph paper. On the left side is the chart template with its charting symbols.

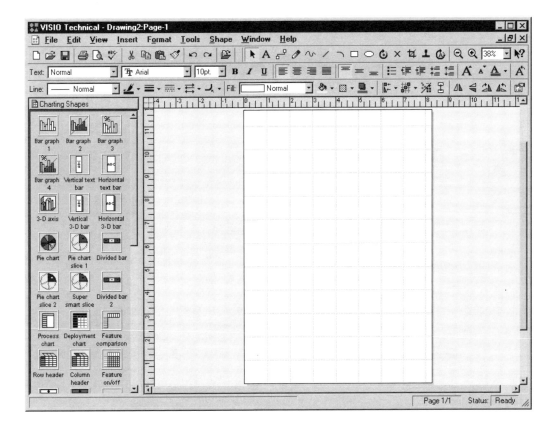

6. Drag the **Bar graph 1** symbol from the stencil to the drawing to get a feel of how Visio works.

7. Enlarge the view of the bar chart to make it easier to see. Right-click the symbol. Move the highlight cursor to **View | 100%**. Click **100%**. Visio increases the size of the symbol.

8. Click the first bar of the chart and type **25**. Visio automatically reduces the height of the bar. Typing a negative number, such as **-90**, makes the bar drop down below the zero line.

9. Use **File | Exit** and click **No** to exit Visio without saving the drawing.

Using a Wizard to Start a New Drawing

1. Start Visio to begin the activity.

2. Notice that Visio displays the Choose a Drawing Template dialog box.

3. Double-click the **Business Diagram** folder to open it.

4. Double-click **Office Layout Wizard** to select it.

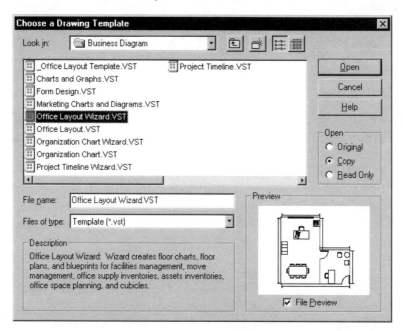

5. Click **OK**. When the Office Layout Wizard welcomes you, click **Next**.

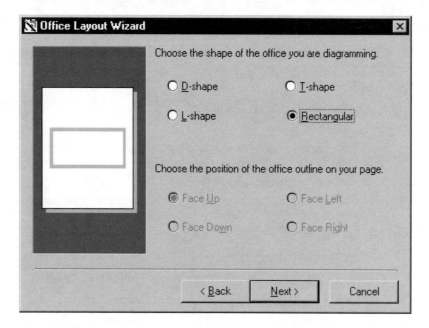

6. Select **Rectangular** when the Office Layout Wizard asks you to select one of four basic floorplan shapes: D-shape, T-shape, L-shape, or rectangular. (When you select one of the other three shapes, you have a choice of direction of the shape faces.) Click **Next**.

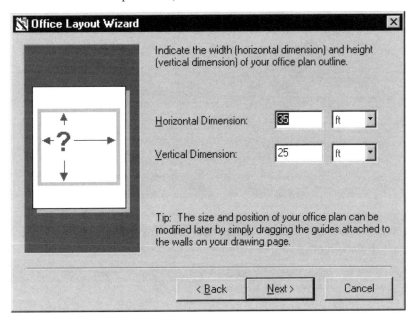

7. Type the length of the walls:
    ➤ Horizontal: **35**
    ➤ Vertical: **25**
    ➤ Select units: **Feet**.

    Click the **Next** button.

    ─────────────

    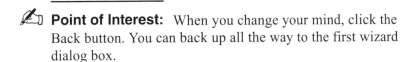 **Point of Interest:** When you change your mind, click the Back button. You can back up all the way to the first wizard dialog box.

8. Click **Finish**. The wizard is finished asking you questions and now goes on to create the floorplan, drawing border, and title block. Click **OK** when told your layout is complete.

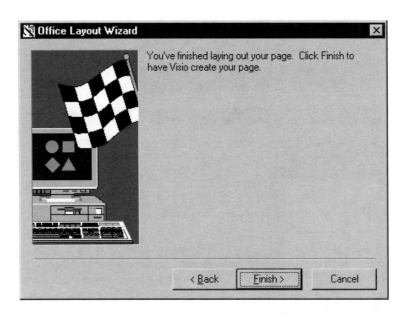

9. You change the location and length of walls by dragging onto the blue guide-lines attached to the walls.

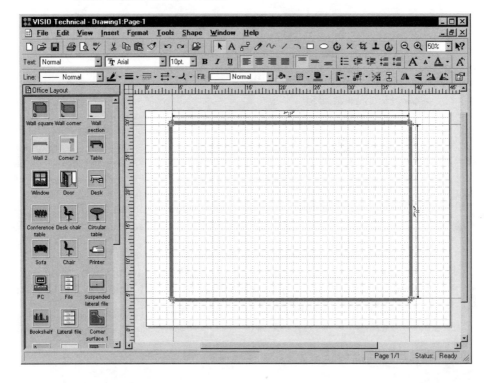

10. Save this drawing with File | Save As as a VST template file (see Module 2, Saving Files) to avoid needing to use the wizard again.

This completes the hands-on activity for opening a new drawing. Do not exit Visio or the drawing since you need both for the next module.

Module 2

Saving Files

File | Save, Save As, Properties

Uses

The **Save** selection of the **File** menu saves the *current* drawing. Saving the drawing to the computer's disk drive lets you work with the drawing again at a later time. If you have more than one drawing loaded into Visio, you must save each one individually.

The current drawing is the *topmost* drawing. When the drawing windows are maximized, the topmost drawing is the one you can see. When the drawing windows are not maximized, the topmost drawing is the one with the highlighted title bar.

When a drawing is new and unnamed, Visio displays the generic name "Drawing1" on the title bar. The first time you save the drawing, Visio asks you to provide a name for the file. With Windows 95 and NT, you type an entire sentence of up to 255 characters long. You may want to limit filenames to eight characters if the drawing will be used by an older version of Visio running on a Windows v3.1 system.

After the first time you save the file by name, **Save** no longer prompts you for the name. Instead, it silently and quickly saves the drawing to disk; the only indication is the hourglass cursor.

When you attempt to exit Visio without saving the drawing, Visio asks, "Save changes to Filename.Vsd?" This gives you one last chance to save the drawing to disk. Click on the **Yes** button.

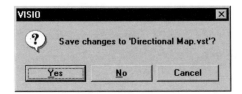

 Point of Interest: You are wise to periodically save (every 15 minutes or so) while you are working on the drawing. That way, you don't lose your work should a power outage or computer breakdown occur. Get into the habit of periodically pressing Ctrl+S to quickly save your work. A large 1MB Visio drawing takes only four seconds to save on a slow Pentium. That's a short time to wait for a large investment in your valuable work.

Save As

When you want to save the file by a different name, use the **File | Save As** command, which displays the **Save As** dialog box. The **Save As** command also lets you save the drawing in file formats read by AutoCAD, Adobe Illustrator, Corel Draw, and earlier versions of Visio.

File Properties

The first time you save a drawing, Visio displays the Properties dialog box. This dialog box has three tabs:

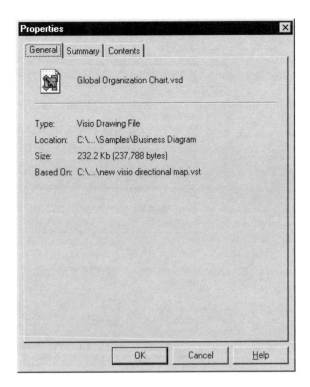

General displays details about the file that comes from the operating system; you cannot edit this data. **Type** describes the class of Visio file, such as "Visio Drawing File." **Location** lists the path to the folder (subdirectory) where the file is located; if the path is too long, an ellipsis (…) helps truncate the full path name. **Size** indicates the size of the file in bytes and KB (kilobytes = 1,024 bytes). **Based On** tells you the name of the VST template file, on which this drawing is based.

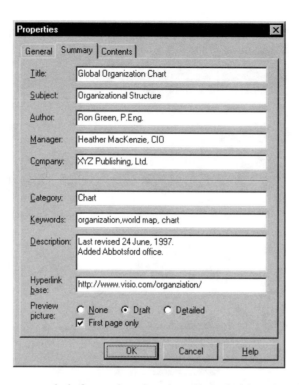

Summary records information that describes the drawing file. In most cases, you can type up to 63 characters in each field. You can change the information at any time. You can use the **Field** command to automatically modify the information. **Title** is a descriptive title of the drawing. **Subject** describes the contents of the drawing. **Author** identifies you. **Manager** is the name of your boss. **Company** is for whom you work—your firm or client. **Category** is a brief description of the drawing, such as map or floorplan. **Keywords** identifies topics related to the file: project name, version number, etc. **Description** allows up to 191 characters. **Hyperlink base** specifies the base URL (uniform resource locator) to be used with the filename. URL is the universal file naming system used by the Internet to identify the location of any file.

Preview picture determines whether a preview image is saved with the drawing (the only reason you would not is to save a bit of disk space). The preview image of the first page appears in the Open dialog box; previews of all pages are available in the Visio Drawing Viewer. **None** doesn't save a preview image. **Draft** saves a preview image of Visio shapes; embedded objects, text, and gradient fills are not displayed. **Detailed** saves the preview image with all objects. **First page only** saves a preview of only the first page.

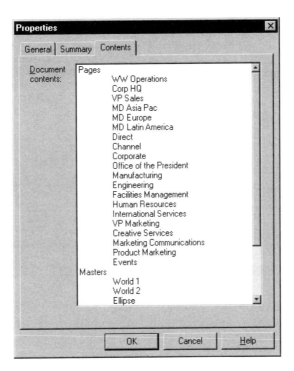

Contents displays a list of the pages and master shapes in the file. A master shape is the shape in the stencil. In the figure above, the drawing contains twenty pages and three master shapes.

To change the file properties at any time, select **File | Properties** from the menu bar.

File Dialog Boxes

All file dialog boxes in Visio allow you to manage files on your computer's disk drives, as well as other drives connected to your computer via a network. To select another folder (subdirectory) or another drive, click the Look In list box.

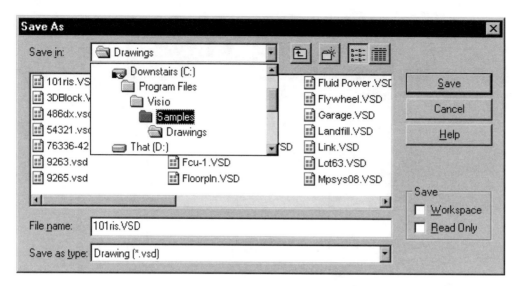

Next to the Look In list box are four icons:

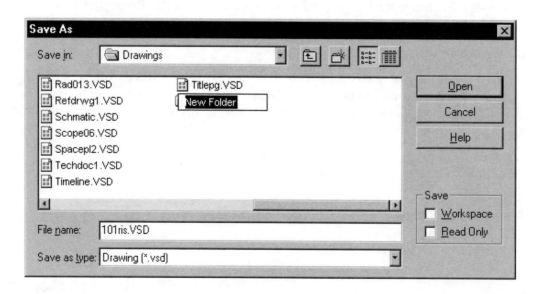

Click the Up One Level button to move up one level in the folder structure.

Click the Create New Folder button to create a new folder. Windows gives the new folder the default name of "New Folder" and allows you to change the folder name to something more meaningful. Like a filename, a folder name can be up to 255 characters long.

Click the List File button to display only filenames, the default. Windows attempts to squeeze in as many filenames as it can.

 Point of Interest: By default, Windows does not display file extensions. To have Windows display the extensions to filenames, start Windows Explorer. From the menu bar, select View | Options. When the Options View dialog box appears, click the check box next to Hide MS-DOS file extensions for file types that are registered. Ensure no check mark appears next to this option. Click OK to dismiss the dialog box.

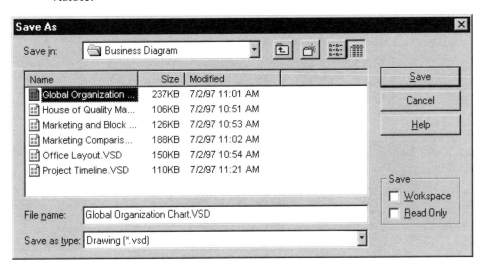 Click the File Details button to display the name, size, and date for every file. Windows displays the details in three columns but does not ensure that all text is visible.

In the figure, the filename "GlobalOrganizationChart.VSD" is truncated to "GlobalOrganization..." To view the entire filename, grab the vertical bar between Name and Size and drag to the right.

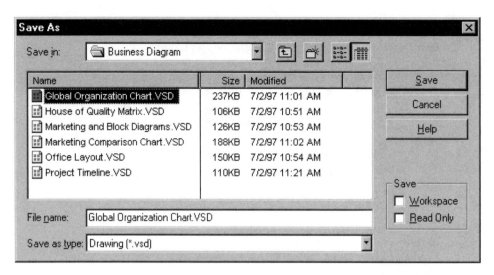

The File Details listing allows you to sort the files in three ways: by filename, size of file, and date last modified. To sort, click the column header:

➤ Click Name to sort the filenames in alphabetical order, from A to Z. Click a second time to reverse the sort, listing filenames from Z to A.

➤ Click Size to sort the files in order of size, smallest to largest. Click a second time to reverse the sort, from largest to smallest (see figure below).

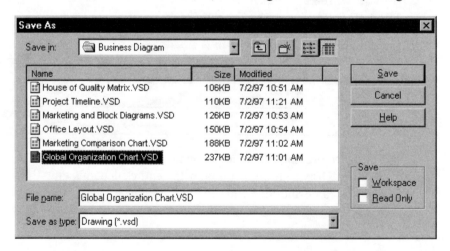

➤ Click Modified to sort the files in order of the data and time stamp, from newest to oldest. This time stamp indicates when the file was last modified. Click a second time to reverse the sort, from oldest to newest.

Right-click a filename to display a floating menu. This menu lets you perform additional file management functions.

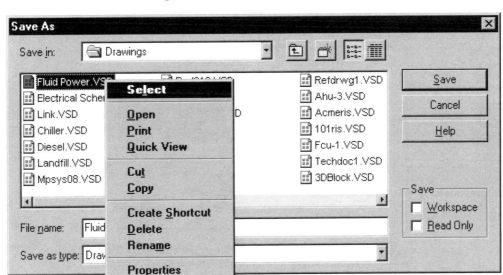

➤ **Select**: Selects the highlighted filename for the File name text box and saves the drawing.

➤ **Open**: Opens the selected filename into Visio; same as using the Open command. The Save As dialog box remains open.

➤ **Print**: Launches a second copy of Visio with the selected filename, and displays the Print dialog box. When printing is complete, the second copy of Visio closes itself. This command fails if the selected drawing is already open in another Visio session.

➤ **QuickView**: Launches the Visio Drawing Viewer (see Module 3, Opening Existing Drawings).

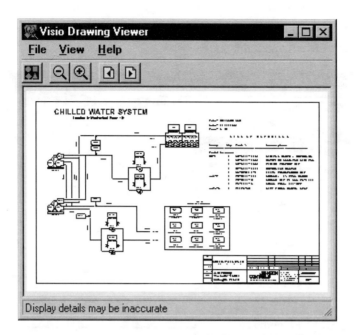

➤ **Cut**: Copies the complete filename, such as "C:\Program Files\VISIO\Samples\Drawings\Chiller.VSD;" to the Windows Clipboard. This can be pasted into another application. What happens in the other application varies: A word processor pastes the Visio drawing; an e-mail program places the file as an attachment. You may need to cancel the Save As dialog box in Visio for the other program to complete its paste operation. Note that the Cut option does not actually cut (erase) the file.

➤ **Copy**: Works the same as the Cut option.

➤ **Create Shortcut**: Creates a shortcut to the selected file in the same folder. You can then drag the shortcut to the desktop.

➤ **Delete**: Sends the file to the Recycle Bin. Windows asks if you are sure.

➤ **Rename**: Allows you to change the name of the file.

➤ **Properties**: Displays a tiled dialog box with the properties of the file.

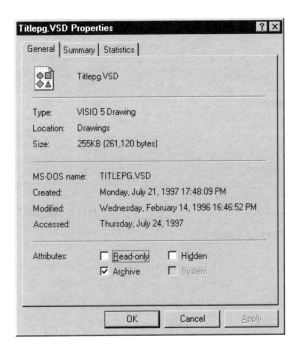

The Save area of the Save As dialog box gives you two options for saving the drawing with extra attributes. These come into effect the next time you load the drawing:

➤ **Workspace**: Saves the drawing, along with the position of windows. This ensures the drawing comes up looking exactly the same way the next time you load it.

➤ **Read Only**: Saves the drawing with the read-only bit set. This means that the next time you load the drawing, you cannot save it. This prevents you (or another user) from making changes to the drawing.

The **Save As Type** list box lets you save the drawing in a large number of different file formats. This is also known as exporting the drawing. (See Module 29, Exporting Drawings.) Visio 5 can export drawings in the following formats:

Visio Formats:	File Extensions
Visio 5.0 Stencil	vss
Visio 5.0 Drawing	vsd
Visio 5.0 Template	vst
Visio 4.x Stencil	vss
Visio 4.x Drawing	vsd
Visio 4.x Template	vst

Internet Formats:	File Extensions
Hypertext Markup Language	htm or html
Drawing Web Format	dwf

Vector Formats:	File Extensions
Adobe Illustrator	ai
AutoCAD Drawing	dwg or dxf
Computer Graphics Metafile	cgm
Encapsulated PostScript	eps
PostScript	ps
Enhanced Metafile	emf
Windows Metafile	wmf
IGES Drawing Format	igs

Raster Formats:	File Extensions:
Graphics Interchange Format	gif
JPEG Format	jpg
Macintosh PICT Format	pct
Portable Networks Graphics	png
Tag Image File Format	tif
Windows Bitmap	bmp or dib
Zsoft PC Paintbrush Bitmap	pcx

Procedures

Before presenting the general procedure for saving the drawing, it is helpful to know about the shortcut keys. The first two save the drawing, the last two trigger an option to save the drawing when it has changed:

Function	Keys	Menu	Toolbar Icon
Save As	none	File │ Save As	
Save	Ctrl+S	File │ Save	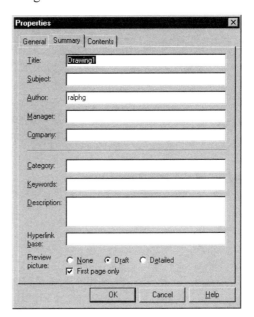
Properties	none	File │ Properties	
Close	Ctrl+F4	File │ Close	
Exit	Alt+F4	File │ Exit	

Saving a Drawing for the First Time

Use the following procedure to save a drawing.

1. Use **File │ Save**.
2. Type the filename.
3. Click the **Save** button.
4. When the Properties dialog box (shown at the right) appears, fill in as much as you like.
5. Click **OK** to dismiss the Properties dialog box.

Saving a Drawing After the First Time

Use the following procedure to save a drawing.

1. Use **File │ Save** (or press **Ctrl+S**).
2. Notice that Visio saves the drawing to disk.

Saving a Drawing As a Template

Use the following procedure to save a drawing as a template file.

1. Use **File │ Save As**.
2. Type the filename.

3. Click on the down-arrow next to the Save as type list box. The list box drops down.

4. Select **Template** (.vst) from the list.

5. Click **Save**. Visio displays the Properties dialog box.

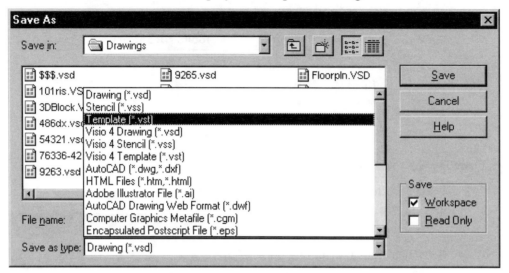

6. Click **OK**. Visio renames the drawing "Drawing1.Vst." The "t" at the end of Vst is a reminder that you are working with a template drawing.

Hands-On Activity

In this activity, you use the Save As function to save the current drawing. Ensure Visio is running and the drawing you created in Module 1 is displayed.

1. Press **Ctrl+S** to save the drawing.

2. Note that Visio displays the Save As dialog box. The default filename, Drawing1, is highlighted since this is first time the drawing is being saved.

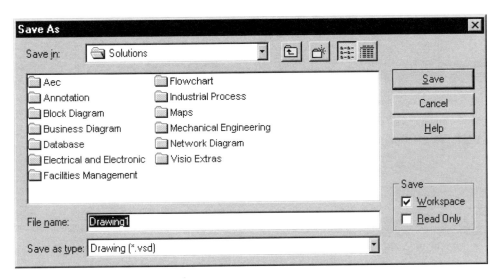

3. Type the filename **Module-2**.

4. Click **Save**.

5. Notice that Visio displays the Properties dialog box. Fill in the information requested by this dialog box:

   ➤ Title: **Module-2**

   ➤ Subject: **Learn Visio 5.0**

   ➤ Author: **your name**

   ➤ Description: **Sample file for practicing the SaveAs command in Module #2.**

   Leave other fields blank.

6. Click **OK**. Visio saves the drawing to disk and changes the name of the drawing to "Module-2.vsd" on the title bar.

7. Press **Alt+F4** to exit Visio.

This completes the hands-on activity for saving the drawing.

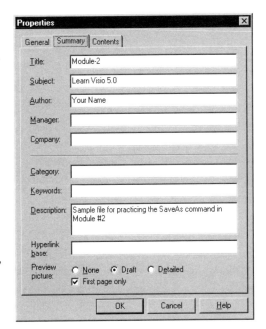

Module 3

Opening Existing Drawings
File | Open, Visio Drawing Viewer

Uses

The **Open** selection of the **File** menu opens an existing Visio template, drawing, stencil, or workspace. By default, Visio displays VST template files in the **Open** dialog box. To open a Visio drawing, instead of a template file, change the **File of Type** to VSD or select **All Visio Files (*.vs*)**; you can open more than one drawing at a time in Visio.

A *template* file (extension VST) is a drawing file that contains custom settings; new drawings are often based on a template file. A *stencil* file (extension VSS) contains shapes that are dragged into the drawing; unlike previous versions of Visio, a drawing no longer requires at least one stencil file. A *workspace* file (extension VSW) records the size and placement of Visio drawing and stencil windows.

You have the option of opening files in three modes. **Original** means that Visio opens the drawing file that you select. **Copy** means an unnamed copy is made from the original file; when you save the drawing, Visio prompts you for a different filename. **Read-only** means that you cannot save the file.

Importing Non-Visio Files

The **Open** command also *imports* files created by other software programs. Visio reads a wide variety of vector, raster, and tagged text formats.

Vector Formats	Extension
ABC FlowCharter	AF2 or AF3
Adobe Illustrator	AI
AutoCAD drawing	DWG
AutoCAD and other CAD softwar	DXF
Computer Graphics Metafile	CGM
CorelDraw drawing	CDR
CorelFLOW chart	CFL
Corel Clipart	CMX
Encapsulated PostScript	EPS
Enhanced Metafile	EMF
Initial Graphics Exchange Specification	IGS
MicroGraphx Designer drawing	DRW
MicroGraphx Designer v6 drawing	DSF
PostScript	PS
Whip Drawing Web Format	DWF
Windows Metafile	WMF

Raster Formats	Extension
Graphics Interchange Format	GIF
JPEG	JPG
Macintosh PICT	PCT
Portable Network Graphics	PNG
Tagged Image File Format	TIFF
Windows Bitmap	BMP or DIB
Z-Soft PC Paintbrush	PCX

Tagged Text Formats	Extension
ASCII text	TXT
Comma-separated value text	CSV

Point of Interest: Be careful: when any program imports a vector file created by a different program, it must translate the data. In some cases, the translation may not be perfect, resulting in some objects being erased; other objects may look different in Visio than in the originating application.

Visio Drawing Viewer

Visio includes an independent utility program that lets you view a Visio drawing without needing to load Visio itself. The advantages are that the viewer loads itself and the drawing image faster than the full Visio program. The image you see in the viewer is the same image displayed by the **Preview** area of the **Open** dialog box.

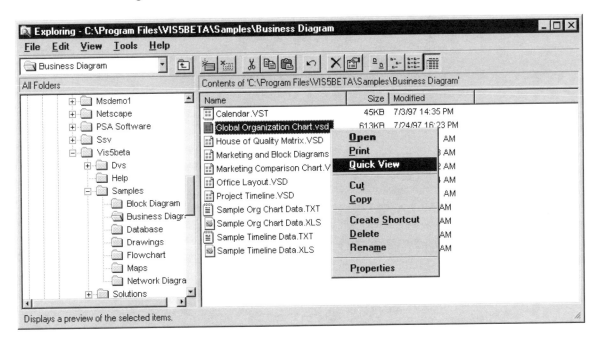

To use the viewer, right click on a Visio file in the Windows Explorer, then select **Quick View**. The Visio Drawing Viewer launches and displays the preview image of the drawing.

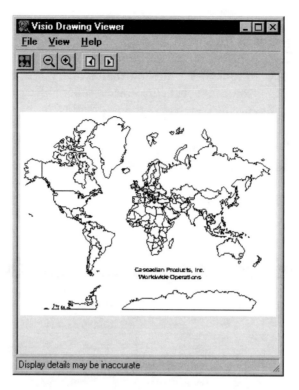

 Open to Visio: Launches Visio and loads the drawing.

Zoom Out: Makes the image smaller so that you can see more of it.

Zoom In: Makes the image larger so that you can see more detail.

Previous Page: When the drawing contains previews of more than just the first page, lets you see the preceding page.

Next Page: Displays the following page, when the drawing contains previews of more than the first page.

To view previews of other drawings, drag their filename from Windows Explorer into the Visio Drawing Viewer. Each time you do so, the viewer launches another copy of itself (unless you turn on the toggle in **View | Replace Window**).

The menu bar provides additional options for the Visio Drawing Viewer:

File: **Open** launches Visio and opens the drawing. **Exit** closes the drawing viewer program.

View: **Toolbar** toggles the display of the toolbar and its icons. **Status Bar** toggles the display of the status bar at the bottom of the viewer. **Replace Window** loads new preview images into the current drawing viewer; when turned off (the default), new preview images are loaded into additional copies of the viewer.

Help: **Help Contents** displays the help file. **About Visio Viewer** displays the viewer's copyright information.

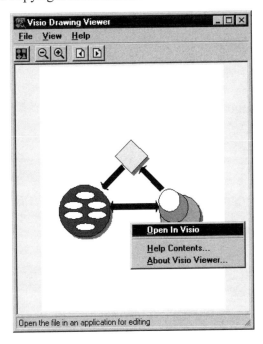

Right-click on the drawing to display the floating menu. **Open In Visio** starts up Visio and loads the drawing. **Help Contents** displays the help file. **About Visio Viewer** displays the viewer's copyright information.

 Point of Interest: The drawing viewer will not display an image under two conditions: (1) the drawing is already open in Visio; and (2) the drawing was saved with the preview option turned off.

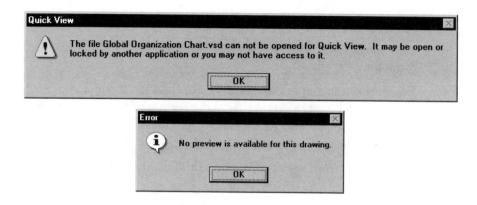

To include a preview image, use the **File | Properties** command, then save the drawing. The preview image, unfortunately, increases the size of the VSD file. For example, the size of the GlobalOrganziationChart.Vsd file grows from 242KB with no preview image to 275KB with a draft preview of the first page, to 627KB with detailed previews of all 20 pages.

Procedures

Before presenting the general procedures for **Open**, it is helpful to know about the shortcut keys. These are:

Function	Keys	Menu	Toolbar Icon
Open	Ctrl+O	File	Open	📂

When you start Visio, it automatically displays the **New** dialog box. Click **Open** to replace the **New** dialog box with the **Open** dialog box.

Opening a Visio Drawing

Use the following procedure to open a Visio drawing:

1. Start Visio.
2. Notice that the **New** dialog box is displayed.
3. Click **Open**.
4. Notice the **Open** dialog box displays names of Visio drawings.
5. Double-click the drawing name.

6. Notice that Visio opens the drawing and associated stencil page.

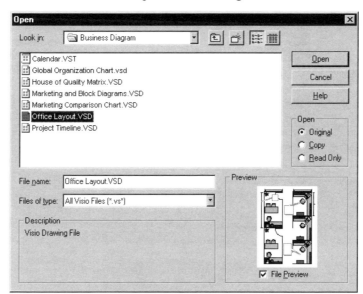

Importing a Non-Visio File

Use the following procedure to open a non-Visio file:

1. Start Visio.
2. Notice that the **New** dialog box is displayed.
3. Click **Open**.
4. Notice the **Open** dialog box displays names of Visio drawings.
5. Click **Files of type**.
6. Select the extension of the file to import.
7. Double-click the filename.
8. Visio opens the file in a drawing without a stencil page.

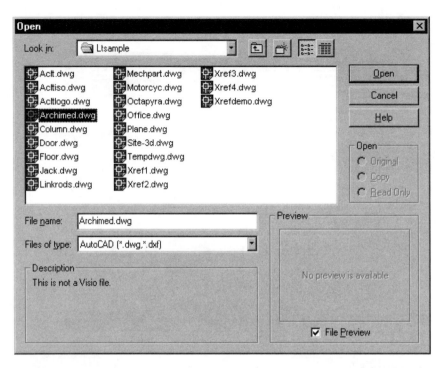

Hands-On Activity

In this activity, you use the Open function. Begin by starting Visio.

1. Click **Open**.
2. Notice the Module-2 filename you saved in Module 2.
3. Double-click the filename.
4. Visio opens the file in a drawing.
5. Press **Alt+F4** to exit Visio.

This completes the hands-on activity for opening a drawing.

Module 4

Setting Up Pages
File | Page Setup, Insert | Page, View | Layer Properties, Edit | Go To)

Uses

The **Page** (on the **Insert** menu) and **Layer Properties** (on the **View** menu) selections prepare a new drawing. To change the properties of an existing page, select **Page Setup** from the **File** menu. The **Go To** selection of the **Edit** menu takes you to another page when your drawing contains more than one page.

"Size" refers to the size and orientation of the page. Visio supports many standard page sizes up to 34" by 44". In addition, you can specify the horizontal and vertical dimensions for any size of page up to 1e19" by 1e19", which is a 1 followed by 19 zeros or ten billion billion inches. That's an area large enough to hold 20,000 copies of our solar system—full size! The orientation is either Portrait (upright) or Landscape (sideways). For a square-shaped page, the orientation doesn't matter.

"Scale" is the relationship between the size of the page and the size of the objects being drawn. For example, when the scale is 1"=10', a shape 1-inch across when printed is displayed as 10 feet wide by Visio's rulers. Visio calculates the scale for you or you can select one of the standard scales:

Scale	*Example*
No Scale	1:1
Metric	1:100
Architectural	3/32" = 1' 0"
Civil Engineering	1" = 10' 0"

Scale	Example
Mechanical Engineering	1/32:1
Custom Scale	1" = 12.5' or 1mm = 254m

"Layers" are a way to separate objects on a page. For example, selecting all objects on a page will not select those objects on a locked layer. When objects are on different layers, changing the layer settings changes how the objects react:

Property	On	Off
Visible	Objects are displayed.	Objects are hidden and do not display.
Print	Objects are printed.	Objects are not printed.
Active	Objects are added to this layer.	Objects are not added to this layer.
Lock	Objects cannot be edited.	Objects are editable.
Snap	Objects snap to other objects.	Objects on other layers cannot snap to objects on this layer.
Glue	Objects glue to other objects.	Objects on other layers cannot glue to objects on this layer.
Color	Objects display in this color.	Objects display their original color.

Visio creates layers automatically when you drag shapes onto the page because many shapes are preassigned to layers. Unlike CAD software, a Visio object can be assigned to more than one layer at a time.

You can create new layers at any time; an option lets you remove unused layers at a later time. That new layer is added to the current page, not all pages in the drawing. In the same way, a new page does not inherit the layers from existing pages in the drawing.

You change the layer settings at any time independently for each layer. For example, if you want to display some text but not print it, you place the text on a layer and turn off the **Print** setting.

The size, scale, and layer settings are not necessary for some drawings. For example, scale often doesn't matter for flow charts and graphs. Unless a drawing is complex, it is easier to draw without worrying about layer settings.

A Visio drawing can have up to 200 pages; by default, a new drawing has just a single page. You can see only one page at a time. A drawing can have many pages, each a different size and scale, and each with many layers. This gives you a great deal of flexibility in displaying and organizing data in your drawing. For example, you can create an entire multipage report with Visio, mixing text and graphics.

Procedures

Before presenting the general procedures for setting up the page, it is helpful to know about the shortcut keys. These are:

Function	Keys	Menu	Toolbar Icon
Size and Scale		File | Page Setup	
Layer Properties		View | Layer Properties	
Add a Page	Shift + F5	Insert | Page	
Go to a Page	F5	Edit | Go To	

Setting the Drawing Size and Scale

Use the following procedure to change the size and scale of a drawing.

1. Select **File | Page Setup** to display the **Drawing Size/Scale** dialog box.

2. Select the **Page Orientation**.

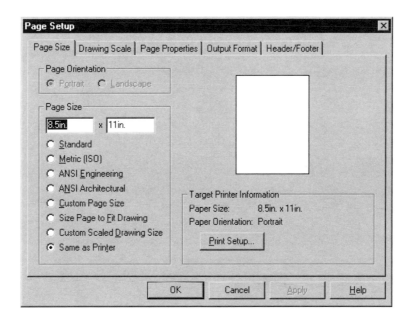

3. Select the **Page Size**.

4. Click the **Drawing Scale** tab to size the drawing.

5. Select **No Scale**. Notice that Visio automatically determines the drawing scale.

6. Click **OK**.

Changing the Drawing Size Interactively

Use the following procedure to make the drawing page larger or smaller:

1. Move the cursor over any edge of the drawing page.

2. Hold down the **Ctrl** key, then drag the edge of the drawing inward (for a smaller page) or outward (for a larger page).

3. Release the **Ctrl** key.

To change the width, grab either side of the page; to change the height, grab either the top or bottom edge; to change both width and height at the same time, grab one of the four corners.

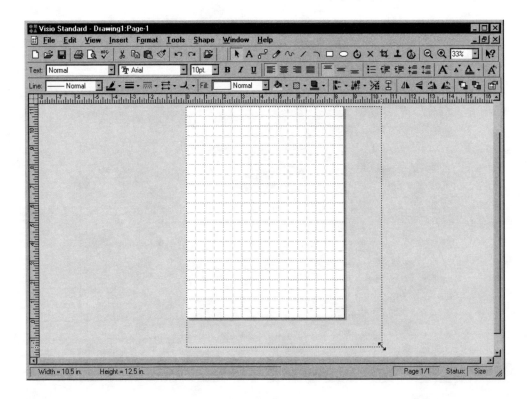

Creating New Layers

Use the following procedure to create a new layer:

1. Select **View | Layer Properties** to display the **Layer Properties** dialog box.
2. Click **New** to create a new layer.

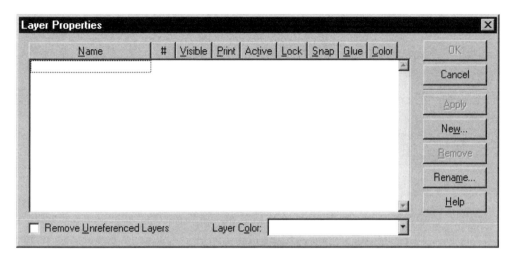

3. Notice that Visio displays the New Layer dialog box.

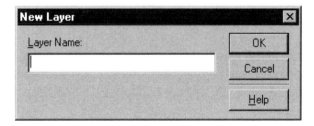

4. Type a name for the layer.
5. Click **OK**. Notice that Visio adds the layer to the list of names.
6. Click **OK**.

Changing Layer Properties

Use the following procedure to change the properties of a layer in a drawing:

1. Select **View | Layer Properties** to display the **Layer Properties** dialog box.

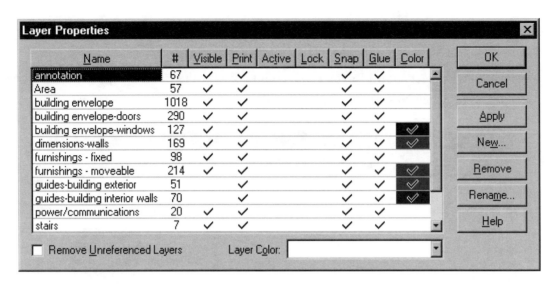

2. Select a layer name.

3. Click **Visible** to change the visibility. A check mark means objects assigned to the layer are displayed; no check mark means the objects are not displayed.

4. Click **Print** to change the printability. A check mark means objects assigned to the layer are printed; no check mark means objects are not printed.

5. Click **Active** to change the layer assignments. A check mark means new objects drawn are assigned to the layer; when more than one layer is made Active, new objects are assigned to all active layers. (Shapes preassigned to a layer go to that layer, not the active layer.)

6. Click **Lock** to change the lock setting. A check mark means objects assigned to the layer cannot be selected or edited. Locked layers cannot change their Visible, Print, Active, Snap, Glue, and Color properties.

7. Click **Snap** to change the snap setting. A check mark means other objects snap to objects assigned to the layer; no check mark means other objects do not snap to objects on this layer.

8. Click **Glue** to change the glue setting. A check mark means other objects glue to objects assigned to the layer; no check mark means other objects do not glue to objects on this layer.

9. Click **Color** to set the color; select the color from **Layer Color**. A check mark means objects assigned to the layer display in the color shown; no check mark means objects take their preassigned color.

10. Click **Visible**, **Print**, **Active**, **Lock**, **Snap**, **Glue**, or **Color** to reverse the property of all layers at once. (This action has no effect on locked layers, ex-

cept the **Lock** property.) The first click changes all layers to the default setting for each property; the second click reverses the setting.

11. Click # to have Visio add up the number of objects assigned to each layer.

12. Click **Apply** to apply changes without leaving the dialog box; click **OK** to apply changes and dismiss the dialog box; click **Cancel** to ignore changes and dismiss the dialog box.

Renaming a Layer

Use the following procedure to change the name of a layer:

1. Select **View | Layer** Properties to display the Layer Properties dialog box.

2. Select a layer name.

3. Click **Rename**.

4. Notice the Rename Layer dialog box.

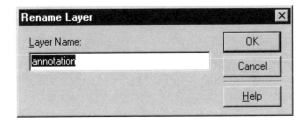

5. Type the new name for the layer.

6. Click **OK** twice.

Removing Layer Names

Use the following procedure to remove an unused layer:

1. Select **View | Layer** Properties to display the Layer Properties dialog box.

2. Select layer name.

3. Click **Remove**.

4. Notice the new Layer Properties dialog box warning.

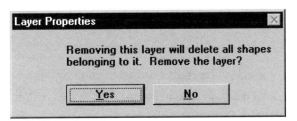

45

5. Click **Yes** to remove layer.

6. To remove all unused layers (layers with no objects assigned to them), click **Remove Unreferenced Layers**.

Creating a New Page

Use the following procedure to create another page:

1. Select **Insert | Page** to display the Page dialog box.

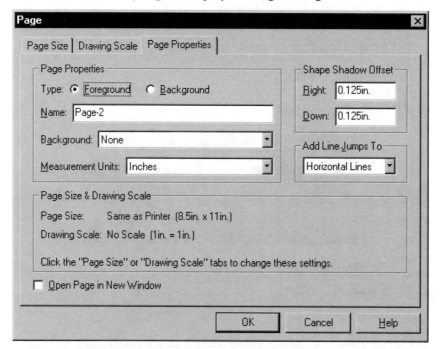

📝 **Point of Interest:** In most cases, the default values provided by Visio are appropriate for a new page and you need only click **OK**.

2. Select **Type**:
   ➤ **Foreground**: shapes are editable.
   ➤ **Background**: shapes are seen but cannot be edited.
3. Type **Name** of up to 31 characters, which is **Page-2** by default.
4. Select the name of a **Background** page to assign to this new page.
5. Select a method of **Measurement** for the rulers. You can have a different measurement system for each page.

6. Type the value for right and down **Shadow Offset**.

 Point of Interest: The Shadow Offset value applies equally to all shapes on this page that have the shadow option turned on. To have the shadow cast up or to the left, use negative values.

7. Click **Open Page** in New Window if you want the page displayed in an independent window; when left unchecked, switching to another page replaces the current page.

8. Click the **Drawing Scale** tab to display the drawing size and scale properties.

9. Click **OK**.

10. Notice that Visio displays a new blank page and states the page number on the title bar.

Moving to Another Page

Use the following procedure to move to different page.

1. Select **Edit | Go To**.

2. Notice the submenu listing the pages in the drawing in the format of Page-*n* (where *n* is the page number).

3. Select a page number.

4. Notice that Visio displays that page, along with the page number on the title bar.

Hands-On Activity

 In this activity, you use the size, scale, and layer functions to set up a new drawing. Begin by starting Visio.

1. Notice the New dialog box.

2. Select **Blank Drawing.VST** to create a new empty document.

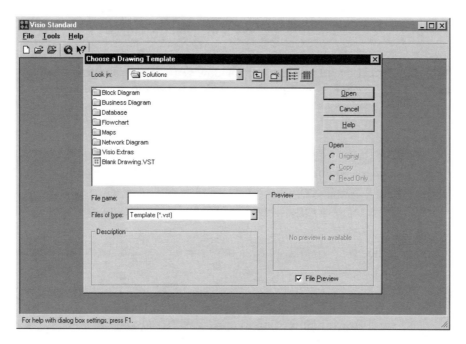

3. Click **Open**.

4. Notice that Visio displays an upright 8-1/2" x 11" page with 1-inch grid lines.

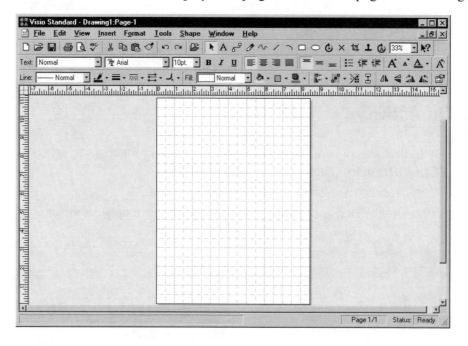

5. Select **File | Page Setup**.

6. Notice the Page Setup dialog box shows the Page Size tab.

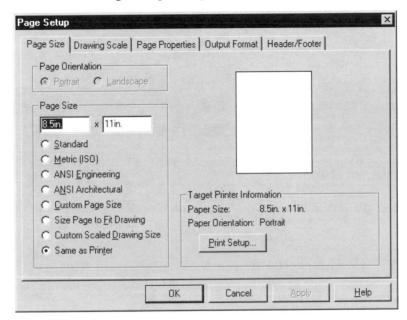

7. Select **Custom Page Size**.

8. Select **Landscape**.

9. Type **11 in x 8.5 in** if Page Size is not those numbers.

10. Click **Apply**. Notice how, underneath the dialog box, Visio changes the page from vertical to horizontal orientation.

11. Click the **Drawing Scale** tab.

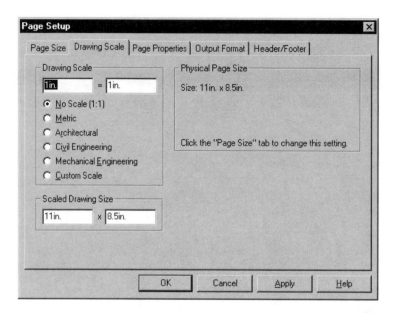

12. Select **Architectural** in the Drawing Scale area of the dialog box.

13. Select Scale of **1/2" = 1'0"**. Notice how Visio adjusts the Drawing Size to 22ft 0in x 17ft 0in.

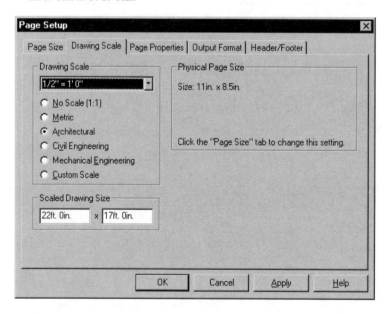

14. Click **OK**.

15. Look at the rulers. Notice that the distance between each grid line represents one scale foot.

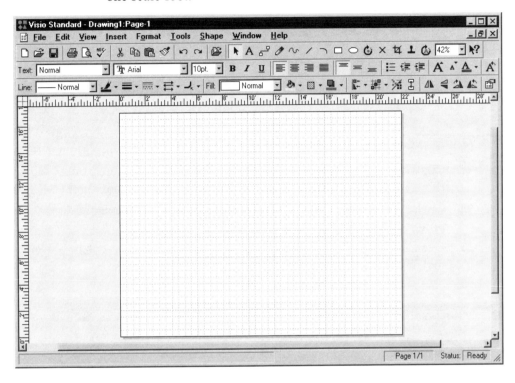

16. Select **View | Layer Properties**. Notice the Layer Properties dialog box has no layer names since this is a brand-new drawing.

17. Click **New** to create a new layer.

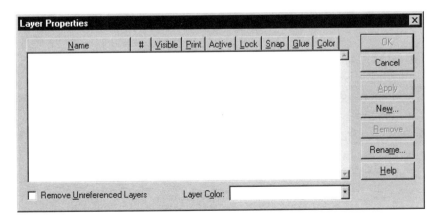

18. Type **Non-printing Text** in Layer Name.
19. Click **OK**.

20. Notice that the Layer Properties dialog box lists layer "Non-printing Text."
21. Click **Layer Color** at the bottom of the dialog box and select **color #7, cyan** (light blue). You may need to scroll though the list of colors to find 7.
22. Click **Print** to turn off printing of this layer. We are making this a non-printing layer.
23. Click **Active** to make this layer the active layer. When you place objects next, they will be on this layer.
24. Click **OK** to dismiss the dialog box.

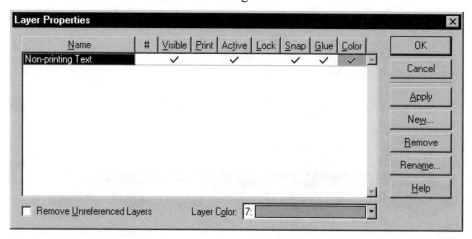

25. Select the **Text** tool [A] (next to the Pointer tool) to place text on the newly created layer.
26. Click near the upper left corner of the page.
27. Type **Drawing for**. The text is placed on layer Non-printing Text and appears in cyan (light blue) color.

28. Notice that Visio enlarges the page and places a boundary box (the dashed lines) around the text.

29. Press **Enter**.

30. Type **Module 4**.

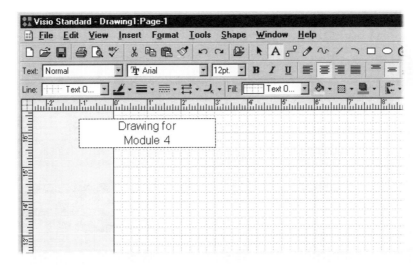

31. Click anywhere else on the page. Notice that Visio returns the page to its original size.

32. Select the **Pointer** tool ▐ ↖ ▐ (looks like an arrow).

33. Click the text. Notice the handles (green squares).

34. Right-click on the text; notice the floating menu.

35. Select **View | Layer Properties**.

36. Click **Active** in the Layer Properties dialog box to turn off the layer.

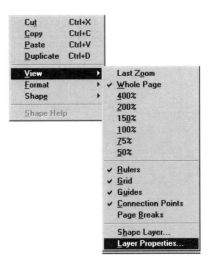

53

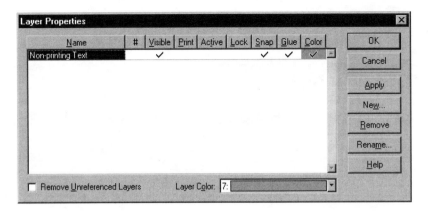

37. Click **OK** to dismiss the dialog box.

38. Use **File | Save As** and save this drawing as Module-4.Vsd.

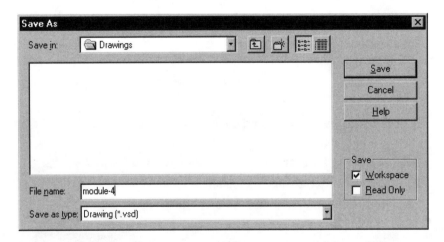

39. Click **Save**.

40. Click **OK** to dismiss the dialog box.

41. Do not exit Visio or the drawing since you need both for the next module.

This completes the hands-on activity for setting up the drawing's size, scale, and layers.

Module 5

Views, Zooms, and Pans

View

Uses

The zoom selections of the **View** menu enlarge and reduce your view of the page within the Visio window. The term "zoom" comes from the camera zoom lens, which brings objects in a scene closer.

The scroll bars along the edge of the Visio window pan the page. "Pan" means to move the view of the page around without changing the zoom level.

Visio enlarges and reduces by a percentage of the actual size. Actual size is "100%." Smaller than actual size is less than 100 percent. For example, 50% is half-size: objects on the page are half their full-drawn size and you see twice as much of the page. The minimum zoom level is 1%, which makes the page 100 times smaller than actual size. This is useful for seeing a very large drawing on the computer's relatively small screen.

Larger than actual size is more than 100 percent. For example, 400% is four times larger: objects in the page are four times their full-drawn size and you see one-quarter of the page. The maximum zoom is 3098%, which makes the page 31 times larger than actual size.

The **View** menu also adjusts the zoom according to the page area. **Page Width** means the page is zoomed so that its width fits the Visio window. **Whole Page** means the entire page is zoomed to fit the window. **Last Zoom** returns to the previous zoom level. **Full Screen** removes the toolbars and other user interface accoutrements, showing the entire drawing. You can draw and edit in full-screen mode using function and control keys. Press **Esc** to return to Visio's normal screen.

By using a combination of the **Ctrl** and **Shift** keys and the mouse buttons, Visio performs quick zooms and real-time pans. When you hold down the **Ctrl** and

Shift keys, Visio displays a magnifying glass cursor to remind you it is now in zoom mode. By holding down together the **Ctrl** and **Shift** keys, you change the view as follows:

View Action	Mouse Action
Zoom in	Click **left** mouse button; each click doubles the zoom percentage.
Zoom out	Click **right** mouse button; each click halves the zoom percentage.
Zoom window	Drag **left** mouse button; windowed area becomes new view.
Pan	Drag **right** mouse button; view pans until button is released.

"Real-time pan" means that Visio pans the page as quickly as you move the mouse. As you hold the **Ctrl** and **Shift** keys, the view pans around the page as you hold down the left mouse button; the cursor changes to the hand cursor. An alternative to zooming in is the "windowed" zoom. By holding the **Ctrl** and **Shift** keys, you draw a rectangle with the right mouse button; that rectangle becomes your new view.

Procedures

Before presenting the general procedures for setting zoom levels, it is helpful to know about the shortcut keys. These are:

Function	Keys	Menu	Toolbar Icon
Actual Size	Ctrl + I	View | Actual Size	
Zoom	F6	View | Zoom	58% ▾
Last Zoom		View | Last Zoom	
Page Width		View | Page Width	
Whole Page	Ctrl + W	View | Whole Page	
Zoom In 200%	Ctrl + Shift + Left-click		🔍
Zoom Out 50%	Ctrl + Shift + Right-click		🔍
Zoom Window	Ctrl + Shift + Right Drag		
Dynamic Pan	Ctrl + Shift + Left Drag		

Zoom In on a Detail

Use the following procedure to enlarge a detail in the page:

1. Click on object to zoom in on it.

2. Select **View** | **Zoom** | **400%**.

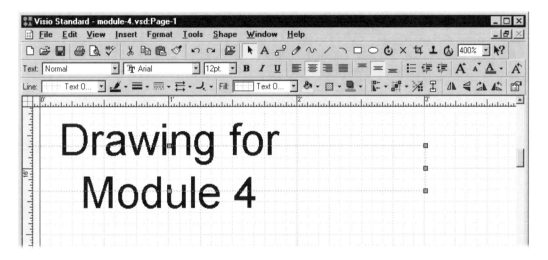

3. Alternatively, press function key **F6**, select **Center Selection in Window**, and select **400%**.

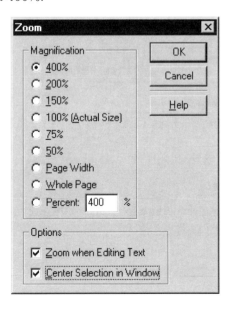

4. Alternatively, right-click the object and select **View | 400%**.

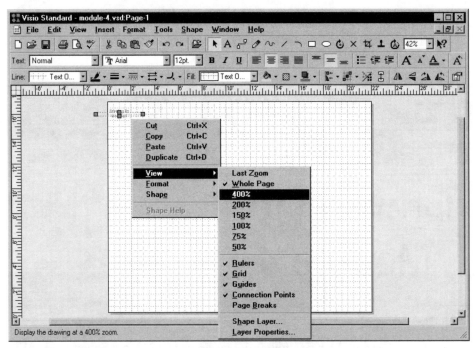

5. Alternatively, select **400%** zoom from the toolbar.

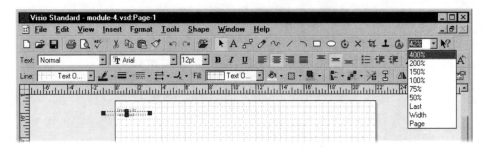

Return to Previous Zoom Level

Use the following procedure to return to the previous zoom level:

1. Select **View | Last Zoom**.

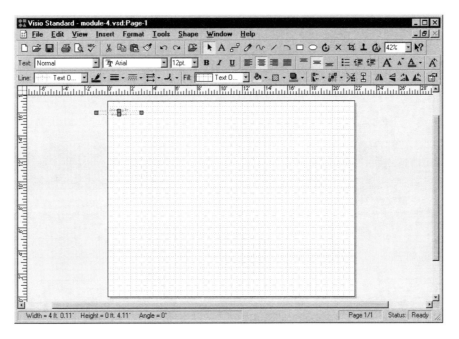

2. Alternatively, right-click and select **View | Last Zoom**.
3. Alternatively, select **Last** zoom from the toolbar.

Hands-On Activity

In this activity, you use the zoom functions. Ensure Visio is running and the Module-4.Vsd drawing is displayed.

1. Click the text to select it.
2. Notice that Visio surrounds the selected text with a dashed green line and green handles. That's Visio's way of giving you feedback. Notice also that Visio tells you the width, height, and rotation angle of the selected object on the status line.

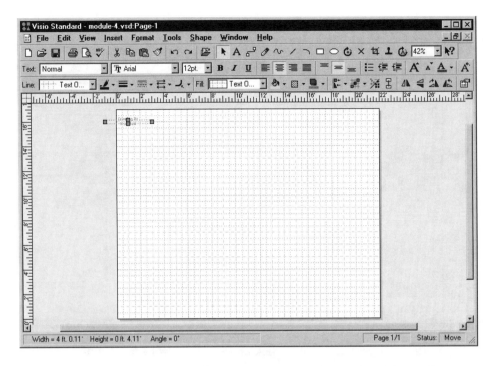

3. Hold down the **Ctrl** and **Shift** keys. Notice that Visio changes the cursor to a magnifying glass with a + sign in it.

4. Move the cursor to the upper left of the text.

5. Press and hold down the left mouse button.

6. With the **Ctrl** and **Shift** keys and the mouse button all held down, move the cursor down and right.

7. Notice how Visio draws a rectangle, which stretches as you move the mouse.

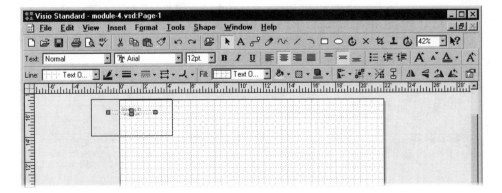

8. Let go of the mouse button. Notice how Visio zooms in and enlarges the text.

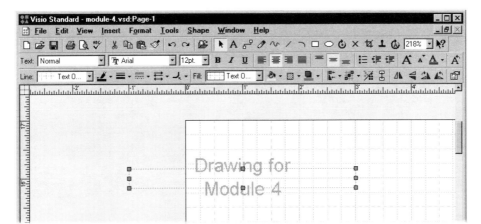

9. Glance at the Zoom Percentage list box. It should read 218% or some number higher than 100.

10. Continuing to hold down the **Ctrl** and **Shift** keys, hold down the right mouse button.

11. Now move the mouse. Notice how the cursor changes from the magnifying glass to an open hand. Notice, too, how the entire drawing moves as you move the mouse.

12. Continuing to hold down the **Ctrl** and **Shift** keys, press the right mouse button repeatedly. As you do, notice how the text becomes smaller as Visio zooms out.

13. Click the **Zoom Percentage** list box and select **Page** to see the entire page again. Alternatively, press **Ctrl+W**.

14. Press **Alt+F4** to exit Visio.

This completes the hands-on activity for zooming in and out of the page.

Module 6

Rulers, Grids, and Guide Lines

View | Ruler, Grid, Guides; Tools | Rulers and Grids;
Tools | Snap and Glue

Uses

The **Ruler**, **Grid**, and **Guides** selections of the **View** menu toggle the display of the rulers, grid lines, and guide lines. The term "toggle" comes from the light switch that turns the lights on and off.

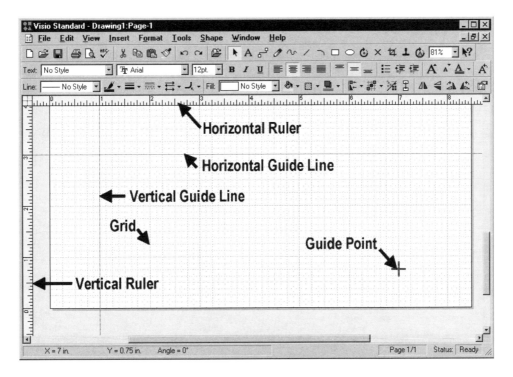

The *ruler* is along the top and left edges of the drawing window. It helps you measure distances. The ruler's measurement system matches that of the units selected for the drawing scale (see Module 4). You select the number of tick marks to display between **Fine**, **Normal**, and **Coarse**. You can move the ruler's origin (zero point) to anywhere in the page. The horizontal ruler's zero point is normally at the upper left corner of the page; the vertical ruler's zero point is normally at the lower left corner of the page.

The *grid* consists of the horizontal and vertical lines on the page. It helps you line up objects as you draw. The grid has four settings: **Normal**, **Fine**, **Coarse**, and **Fixed**. You specify how far apart the grid lines are. You can move the origin of the grid to align with an object on the page.

The *guide line* is like a customizable grid line. You create a guide line by dragging the guide line from the ruler (horizontal or vertical) over onto the page. Guide lines help you align objects.

A *guide point* is a small cross marking a point. You create it by dragging from the intersection of the two rulers. A page can have many guide lines and guide points, although too many obscure the drawing.

Snap is the ability of Visio to cause objects to line up. Snap makes it easier to create an accurate drawing. You can have Visio snap to the ruler's tick marks, to the grid, to objects, and to guides. You adjust the snap's strength. For example, if the cursor is within 5 pixels of a grid line, then the snap takes place. Other default values are within 3 pixels of a ruler tick mark and 8 pixels of a guide line or point.

Glue is the ability of objects in Visio to stay together when moved or stretched.

You can change the ruler, grid, ruler, snap, and glue settings at any time without affecting objects already placed in the drawing.

 Point of Interest: **Fine**, **Normal**, and **Coarse** are relative terms. In *normal* spacing, the ruler and grid spacing change as you zoom in or out; *fine* displays twice as many grid lines as normal spacing, while *coarse* displays half as many grid lines. **Fixed** sets the grid; when you zoom in and out, the spacing between grid lines doesn't change.

Procedures

Before presenting the general procedures for ruler, grid, ruler, snap, and glue settings, it is helpful to know about the shortcut keys. "Toggle" means to turn on and off. These are:

Function	Keys	Menu	Toolbar Icon
Glue toggle	F9		⊞
Snap toggle	Shift + F9		⊞
Ruler toggle		View | Rulers	
Guides toggle		View | Guides	⊞
Grid toggle		View | Grid	⊞
Ruler and Grid dialog		Tools | Ruler and Grid	
Snap and Glue dialog	Alt + F9	Tools | Snap and Glue	

Setting the Snap and Glue

Use the following procedure to set the snap and glue:

1. Select **Tools | Snap & Glue**.
2. Notice the Snap & Glue dialog box.

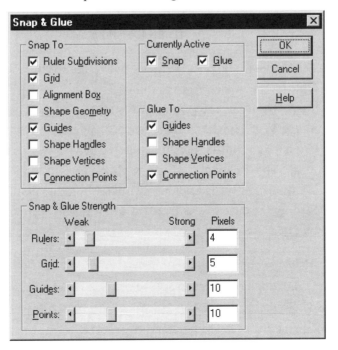

65

3. Notice that the Currently Active section displays the current toggle setting of Snap and Glue.

4. Select the items you want Visio to snap to in the Snap To section:
   - ➤ **Ruler Subdivisions**: the tick marks on the rulers.
   - ➤ **Grid**: the grid lines.
   - ➤ **Alignment Box**: the black rectangle that surrounds a shape while the shape is being moved.
   - ➤ **Shape Geometry**: vertices and endpoints of lines and arcs.
   - ➤ **Guides**: the guide lines and points.
   - ➤ **Shape Handles**: the green squares surrounding a shape.
   - ➤ **Shape Vertices**: the corners of the green dotted line surrounding a shape.
   - ➤ **Connection Points**: the blue x where a shape connects with another shape.

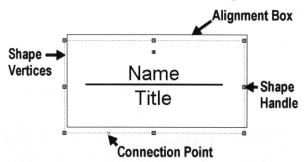

5. Select the items you want Visio to glue to in the Glue To section:
   - ➤ **Guides**: the guide lines and points.
   - ➤ **Shape Handles**: the green squares surrounding an object.
   - ➤ **Shape Vertices**: the diamond shape handle between two segments.
   - ➤ **Connection Points**: the point where objects are connected together.

6. Move the slider bars in the Snap & Glue Strength section:
   - ➤ **Weak**: the cursor is as close as 1 pixel from the point.
   - ➤ **Strong**: the cursor is as far away as 30 pixels from the point.

7. Click **OK**.

Setting the Ruler and Grid

Use the following procedure to set the ruler and grid.

1. Select **Tools | Ruler & Grid**.

2. Notice the Ruler & Grid dialog box.

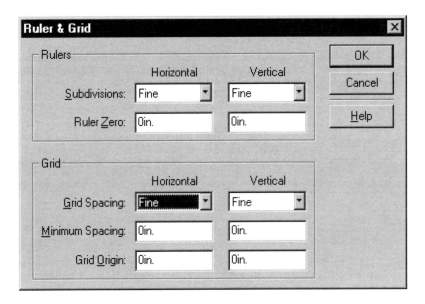

3. Select settings for the ruler from the Rulers section:

   ➤ **Subdivisions**: select from fine, normal, or coarse.

   ➤ **Ruler Zero**: specify a horizontal and vertical scale distance from the default point.

4. Select settings for the grid from the Grid section:

   ➤ **Grid Spacing**: select from fine, normal, coarse, or fixed spacing. Fine spacing displays twice as many grid lines as the normal setting; coarse displays half as many grid lines as normal spacing.

   ➤ **Minimum Spacing**: for fine, normal, and coarse, specify the minimum distance between grid lines. For fixed, specify the distance between grid lines. Visio always displays the fixed grid lines, no matter the zoom level.

   ➤ **Grid Origin**: specify a horizontal and vertical grid distance from the default origin.

5. Click **OK**.

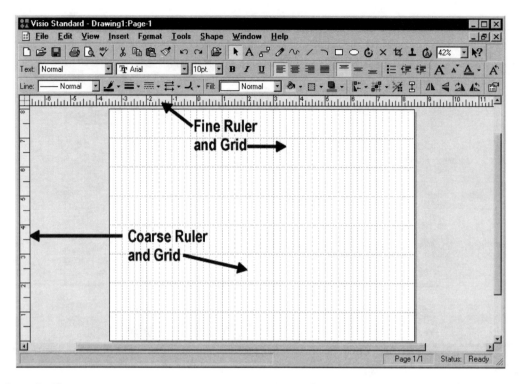

Toggling the Snap

Use the following procedure to turn off and on the snap:

1. Press function key **Shift+F9**.
2. Alternatively, press the **Snap** button on the toolbar.

Toggling the Glue

Use the following procedure to turn off and on the glue:

1. Press function key **F9**.
2. Alternatively, press the **Glue** button on the toolbar.

Creating a Guide Line

Use the following procedure to create one or more guide lines:

1. Move the cursor to the horizontal or vertical ruler. Notice that you can create guide lines during any command.
2. Click the mouse button, then drag into the drawing.
3. Notice the blue line moving with the cursor; this is the guide line.

4. Release the mouse button once the guide line is in position. Notice that the guide line turns green, indicating it is selected.

 Point of Interest: You can drag as many guide lines into the drawing as you require. Dragging from the horizontal ruler results in a horizontal guide line; dragging from the vertical ruler results in a vertical guide line. Guide lines respect the setting of snap, grid, ruler, connection points, etc. To position the guide line accurately, make use of those positioning aids. An unselected guide line is colored blue.

Repositioning a Guide Line

Use the following procedure to reposition a guide line:

1. To reposition a guide line, click the **Pointer Tool** icon.
2. Select the guide line. Notice that its color changes from blue to green.
3. Drag the guideline to a new position.
4. To move more than one guide line at a time, hold down the **Shift** key while selecting guide lines. The first guide line you select turns green; the additional guide lines you select turn cyan (light blue).

Removing a Guide Line

Use the following procedure to remove one or more guide lines:

1. To remove a guide line, click the **Pointer Tool** icon.
2. Select the guide line by clicking it. Notice that it turns green.
3. Press the **Delete** key. Visio erases the guide line.
4. To erase more than one guide line at a time, hold down the **Shift** key while selecting guide lines, then press the **Delete** key.

Creating and Removing Guide Points

Use the following procedure to create and remove one or more guide points:

1. Move the cursor to the intersection of the rulers (upper left corner of Visio's drawing area).
2. Click and drag into the drawing. Notice the two blue lines moving with the cursor.
3. Let go of the mouse button when the guide point is in position. The guide point looks like a small green plus-sign.

4. To erase the guide point, select it and press the **Delete** key.

Toggling the Display of Guide Lines and Points

Use the following procedure to turn off and on the display of guide lines and points:

1. Select **View | Guides**. The check mark indicates guide display is on; no check mark means guides are not displayed.
2. Alternatively, press the **Guides** button on the toolbar.
3. Alternatively, create a new layer and place guides on that layer. Toggle the visibility of the guide lines and points by turning the visibility of that layer off and on.

Toggling the Grid Display

Use the following procedure to turn off and on the display of grid lines:

1. Select **View | Grid**. The check mark indicates grid display is on; no check mark means grid is not displayed.
2. Alternatively, press the **Grid** button on the toolbar.

Relocating the Ruler Origin

Use the following procedure to move thezero setting of the ruler:

1. Move the cursor over the vertical or horizontal ruler.
2. Hold down the **Ctrl** key.
3. Hold down the left mouse button.
4. Drag into the drawing. Notice that a black line moves with the mouse into the drawing.
5. Let go of the **Ctrl** key and mouse button. Notice that the zero point on the ruler has moved.
6. To reset the ruler's zero point back to its default position, double-click the other ruler.

To change both rulers at the same time, drag from their intersection at the upper left corner. To reset both rulers at the same time, double-click the intersection point.

Hands-On Activity

In this activity, you use the rule and grid functions. Ensure Visio is running and start a new drawing.

1. Select **Tools | Ruler & Grid**. Notice the Ruler & Grid dialog box. All settings are either Fine or 0.

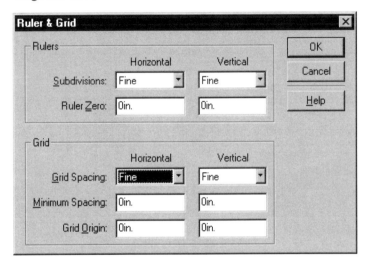

2. Select **Coarse** for the Horizontal and Vertical ruler Subdivisions.

3. Select **Fine** for the Horizontal and Vertical Grid Spacing.

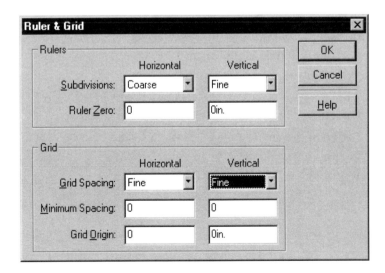

5. Click **OK**.

6. Notice the coarse ruler tick mark spacing and the fine grid line spacing.

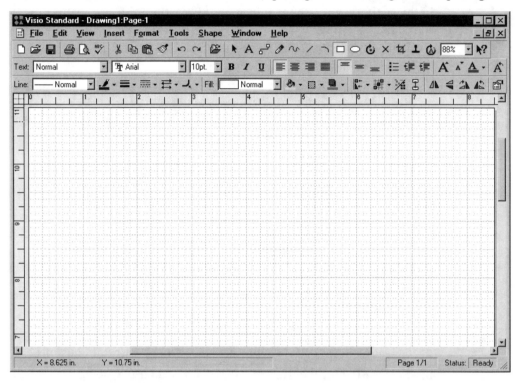

7. To place a vertical guide line, move the cursor to the vertical ruler (located at the left edge of the Visio drawing area). Notice that the cursor becomes a horizontal double-ended arrow.

8. Hold down the left mouse button.

9. Drag the mouse into the drawing. Notice the blue vertical line moving with the mouse.

10. Let go of the mouse button. Notice that the guide line changes from blue to green.

11. Exit Visio by pressing **Alt+F4**.

This completes the hands-on activity for setting the ruler and grid.

Module 7

Opening Existing Stencils

File | Stencils

Uses

The **Stencils** selection of the **File** menu lets you open one or more stencil files, which are used to place shapes in the page. *Stencils* are similar to the green plastic symbol templates used by drafters to quickly draw commonly used shapes.

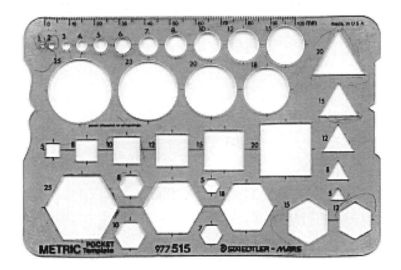

In Visio, stencils contain predrawn objects, called *shapes*. Visio shapes have more intelligence than the shapes cut out of the green plastic. The shapes know their logical connection point and are automatically assigned a layer name. Unlike the plastic stencil shapes, Visio shapes are easy to resize, easy to modify, and are in full color.

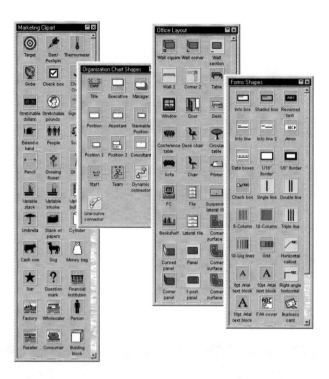

Visio Standard comes with 39 stencil files; Visio Professional has 90 stencil files; and Visio Technical includes 118 stencil files (files with extension .VSS found in the \Solutions subdirectory). Each stencil file typically includes 25 to 35 shapes. That means you have as many as 3,500 shapes at your disposal. Additional specialized stencil files are available from Visio Corp. and third-party developers. To help you find stencil files, whether on your computer or your firm's network or on the Internet, Visio includes the Shape Explorer.

You use a shape in two steps: (1) open the stencil file; and (2) drag the shape from the stencil to the page. In this module, you learn the first step, opening the stencil file. Actually, the shape in the stencil is called the *master shape* because Visio makes a copy of the master when you appear to drag the shape into the drawing.

Procedures

Before presenting the general procedures for opening a stencil file, it is helpful to know about the shortcut keys. These are:

Function	Menu	Toolbar Icon
Open stencil file	File | Stencils | Open Stencil	
Open blank stencil	File | Stencils | Blank Stencil	
Shape Explorer	File | Stencils | Shape Explorer	

Opening a Stencil File

Use the following procedure to open a stencil file:

1. Select **File | Stencils | Block Diagram | Basic Shapes**. Notice how Visio groups the stencil files in this menu selection.

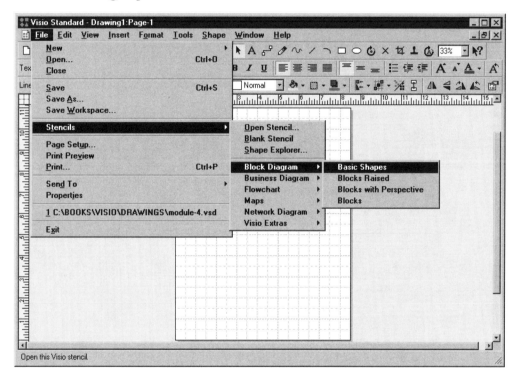

2. Notice how Visio resizes the page window to accommodate the stencil file.

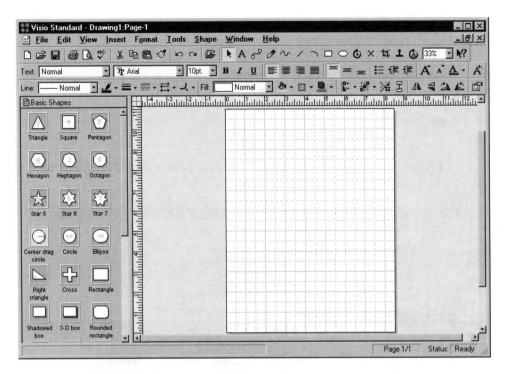

3. Notice the shapes have two background colors:
   ➤ Gray indicates a regular shape.
   ➤ Yellow indicates a "SmartShape," a shape with extra intelligence hidden inside it.
4. You load more than one stencil file by repeating the procedure and selecting a different stencil name.

Adjusting the Stencil Window

Use the following procedure to change a stencil window:

1. Click the icon on the stencil's title bar.
2. Notice the options:
   ➤ **Activate**: Brings the stencil to the foreground when more than one stencil file is loaded.
   ➤ **Float**: Makes the stencil independent of the Visio page window.
   ➤ **Switch Sides**: Moves stencil to right side of Visio page window.
   ➤ **Close**: Closes (removes) the stencil.

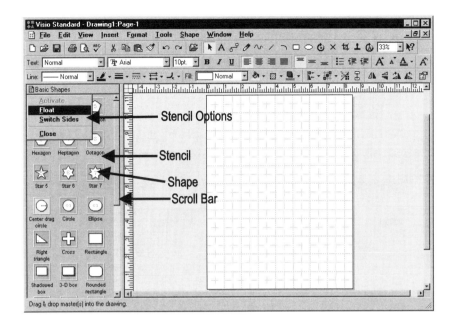

3. Drag the sroll bar to see more shapes in the lower part of the stencil.

4. Right-click on a shape in the stencil.

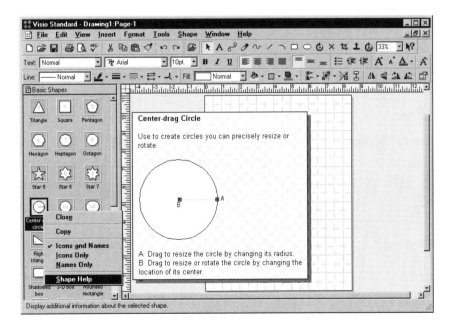

5. Notice the options:

> ➤ **Close**: Closes (removes) the stencil.

> ➤ **Copy**: Copies the shape to the Windows Clipboard in picture (WMF) format.

> ➤ **Icons and Names**: Displays shapes as icons and their names (the default)

> ➤ **Icons Only**: Displays the shapes as icons without their names.

> ➤ **Names Only**: Displays the shapes by their names only.

> ➤ **Shape Help**: Displays a window describing how to use the shape.

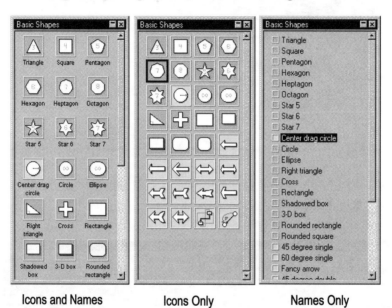

Icons and Names   Icons Only   Names Only

Using Shape Explorer

Use the following procedure to explore shapes:

1. From the menu, select **File | Stencils | Shape Explorer**. Notice the Shape Explorer that Visio launches.

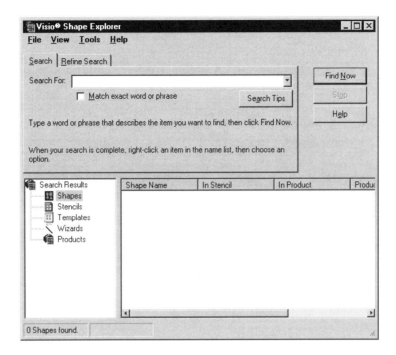

2. Type a word or phrase in the **Search For** text box.

 Point of Interest: Shape Explorer has three primary methods for controlling its search:

➤ **Match Exact Word Or Phrase**: Shape Explorer searches for an exact match to the words you type. This is equivalent to using quotation marks, such as "Blocks with Perspective." When you type a single word, Shape Explorer searches for items that contain only that word. For example, typing "block" returns "Block Diagram" but not "Blocks Raised."

When this option is turned off, Shape Explorer searches for items that have those words, such as "blocks," "with," and "perspective." This could include "Block Diagram," "Blocks Raised," and "Blocks with Perspective."

➤ **AND**: The AND operator narrows the search by forcing Shape Explorer to only find items that contain both words. For example, "Blocks AND Raised" only returns "Blocks Raised."

➤ **OR**: The OR operator broadens the search by forcing Shape Explorer to find all items that contain any of the words. For example, "Blocks OR

Raised" returns "Blocks Raised," "Blocks with Perspective," and "Raised Ck. Top."

3. Click the **Find** button. Notice that Shape Explorer spends a minute or two searching through Visio's .MDB database files.

4. When done, Shape Explorer displays a summary of its results in the **Your Search Found** window. The number in the parentheses indicates the number of items found. For example, Shapes (4) means four shapes were found.

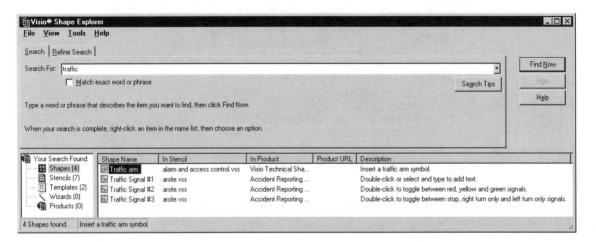

5. Click on each of the five category names: **Shapes**, **Stencils**, **Templates**, **Wizards**, and **Products**. Notice how the right-hand window displays the details for each category:

> **Name**: The full name of the shape, stencil, template, wizard, or product.

> **In**: The location of the item, whether in a stencil file, template file, or product filename. For example, shapes are found in stencil VSS files.

> **Product URL**: When the item is found on the Internet, then the URL is displayed. URL is short for "uniform resource locator" and is the universal file naming system for the Internet. An example of a URL is the familiar Web address, such as http://www.visio.com.

> **Description**: A one-sentence description of the item.

6. Right-click on an item to see the list of options:

> **Browse Info @ www.Visio.com**: Launches your computer's Web browser.

> **Properties**: Displays a dialog box with some additional information about the selected item.

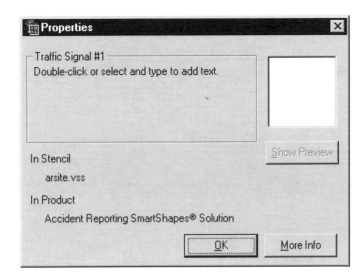

7. Unfortunately, you cannot drag the shape from Shape Explorer into your drawing. Instead, memorize the filename and use the **Open** command to load the file.

Hands-On Activity

In this activity, you open two stencil files.

1. Select **File | Stencils | Business Diagram | Office Layout Shapes** from the menu bar.
2. Notice that Visio resizes the page window to accommodate the Office Layout stencil file.
3. Drag the stencil's scroll bar to look at the 35 shapes in the stencil.
4. Open a second stencil by selecting **File | Stencils | Business Diagram | Marketing Diagram**.
5. Notice how the newly opened stencil covers up the first stencil. In case you don't see it, the title bar of the Office Layout is at the bottom of the Visio window.

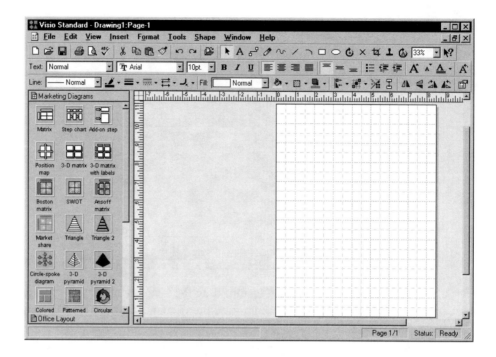

6. To bring the first stencil back into view, click on the **Office Layout** title bar. Notice how it appears to slide up, covering over the Marketing Diagram stencil.

7. To see both stencils at the same time, right-click on the stencil's title bar. Visio displays a floating menu.

8. Select **Switch Sides**. Notice how both stencils are visible, one on either side of the drawing area.

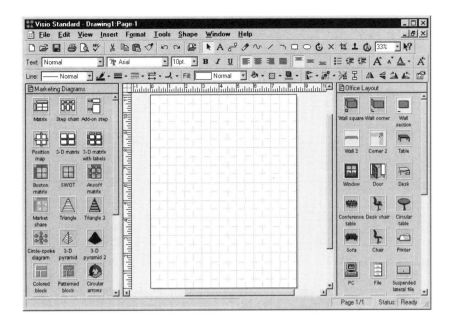

9. To make more room for the drawing area, you have three options:

   ➤ Drag the side of the stencil to make it narrower.

   ➤ Right-click on any master shape. Select Icons Only or Names Only from the floating menu.

   ➤ Right-click the title bar and select Float. Notice that the stencil becomes an independent window and can be placed anywhere on your screen, even outside of Visio.

10. Exit Visio with **Alt+F4**.

This completes the hands-on activity for opening and positioning a stencil file.

Module 8

Dragging Master Shapes into the Drawing

Uses

To use a shape, you drag it from the stencil to the drawing. There are no menu selections, toolbar icons, or shortcut keys for placing shapes in the drawing.

Procedures

Use the following procedure to drag a shape in the drawing. Ensure Visio is running with at least one stencil file open and a drawing page displayed.

1. Move the cursor over a shape in the stencil.

2. Drag the shape to the drawing. You "drag" by holding down the left mouse button, then moving the shape into the drawing, and letting go of the mouse button.

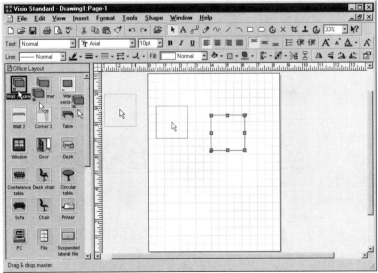

Point of Interest: You may drag the same shape from the stencil to the page many times. A Visio stencil does not "run out" of shapes.

Hands-On Activity

In this activity, you start to create an office drawing by dragging shapes onto the page. Ensure Visio is running and is displaying the Office Layout stencil file.

1. Move the cursor over to **Wall square** master shape.

2. Click left mouse button.

3. Notice that Visio highlights the shape by placing a heavy black border around it. Notice also the hint on the status line: "Stretch rectangular wall to desired dimensions."

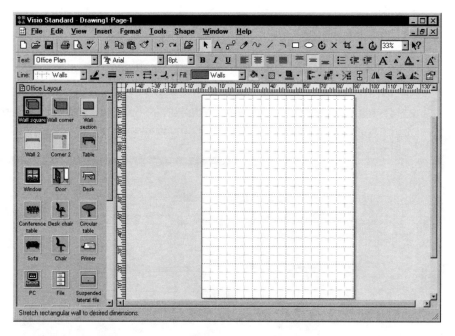

4. Right-click on **Wall square**.

5. Select **Shape Help**.

6. Read the help provided for Wall Square:

Wall Square

Use to form the outer walls of a room or building. Position the walls outside the room's perimeter.
To associate inventory number and owner information with the shape, right-click the shape, then choose Properties.
To set a minimum wall thickness for all the walls in a drawing, make sure no shapes are selected, right-click the drawing page, then choose Properties.
To see dimensions as you resize, connect a Dimension Line shape to this shape. To connect a Dimension Line shape to this shape, glue each endpoint on the Dimension Line shape to separate connection points × on this shape.
Control handles provide other shape actions. To see what a control handle ▧ on a selected shape does, pause the pointer over the handle.

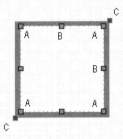

A   Drag to resize the wall square proportionally.
B   Drag to lengthen the wall square in one direction.
C   Drag horizontally or vertically to thicken one wall. Drag diagonally to thicken two walls at once.

7. Click on the help window to dismiss it.

8. Drag the **Wall square** shape to the drawing.

9. Notice that the wall has the same green squares (handles) as were shown in the help box.

10. Right-click the shape to display a menu with the following commands:

➤ **Cut** cuts the shape from the drawing and places it in the Windows Clipboard.

➤ **Copy** copies the shape to the Clipboard. It is available to Visio and other Windows applications in five different formats: Visio 5 drawing, Visio drawing data, Picture (WMF), Enhanced Picture (EMF), and ANSI text of the shape name. See Module 12.

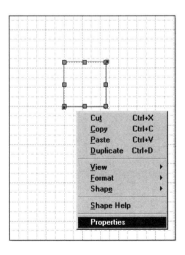

➤ **Paste** pastes whatever is currently in the Clipboard into the center of the drawing.

➤ **Duplicate** makes a copy of the selected shape, slightly offset.

➤ **View** lets you change the zoom level and toggle the display of drawing aids, such as grids and rulers.

➤ **Format** lets you change the formatting of the lines and text that make up the shape. See Modules 13 and 14.

➤ **Shape** lets you change the position of the shape, such as flipping it. See Module 9.

➤ **Shape Help** displays the same help box we saw earlier.

➤ **Properties** varies with the shape you select. For this particular wall shape, it displays the **Custom Properties** dialog box that lets you type an inventory number and name of owner. Filling in this information is completely optional; the data can be extracted to a database to help you create an inventory listing. Other shapes have other custom properties. For example, the Cray computer shape has text fields for ID, Location, Manufacturer, Product Name, Model Number, and Description. In some cases, the fields are filled in for you, but you can edit them. See Module 34.

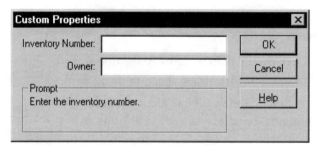

11. Exit Visio with **Alt+F4.**

This completes the hands-on activity for placing a shape in the drawing.

Module 9

Sizing and Positioning
Shape | Size and Position (Rotate, Flip, Reverse Ends, Bring to Front, Send to Back); Tools | Center Drawing

Uses

The **Size and Position** selection of the **Shape** menu is used to change the size and position of the selected shape. More often, you will use the shape's handles (the green squares) to size and position the shape. The term *handle* comes from handles on suitcases and mugs that help you carry and hold them. Via the handles, you change the size of a shape, stretch the shape, move the shape, duplicate the shape, change its characteristics, and erase the shape, as follows:

➤ Inside corner handles: changes the size proportionately.

➤ Inside mid handles: stretches vertically or horizontally.

➤ Outside corner handles: thicken the walls.

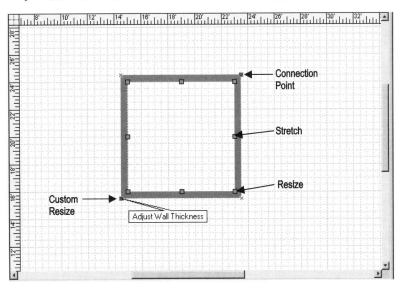

The small blue x markers indicate connection points, where other shapes automatically connect. Other shapes have handles that perform other actions. When you see a small padlock, it means the shape cannot be resized.

The **Rotate** selection of the **Shape** menu rotates the shape at right angles (90 degrees) each time you use the command. **Rotate Left** rotates the shape 90 degrees counterclockwise; **Rotate Right** rotates the shape 90 degrees clockwise. In addition, Visio has the **Rotation Tool** that lets you rotate the shape by any angle.

The **Flip** selections of the **Shape** menu transpose the shape. **Flip Horizontal** transposes the left and right halves, while **Flip Vertical** reverses the top and bottom halves.

The **Reverse Ends** selection of the **Shape** menu mirrors the shape about its center point. It is like using the **Flip Vertical** and **Flip Horizontal** functions at the same time.

The **Bring to Front** selection of the **Shape** menu moves a shape in front of an overlapping shape; the **Send to Back** selection moves the overlapping shape to behind the underlying shape. When three or more shapes overlap, the **Bring Forward** and **Send Backward** move the selected shape by one shape at a time.

The **Center Drawing** selection of the **Tools** menu centers the entire drawing on the page.

Procedures

Before presenting the general procedures for sizing and positioning it is helpful to know about the shortcut keys. These are:

Function	Keys	Menu	Toolbar Icon
Size and Position dialog box		Shape | Size and Position	
Flip horizontal	Ctrl+H	Shape | Flip Horizontal	
Flip vertical	Ctrl+J	Shape | Flip Vertical	
Free rotation	Ctrl+0	Shape | Size and Position	
Rotate left	Ctrl+L	Shape | Rotate Left	
Rotate right	Ctrl+R	Shape | Rotate Right	
Reverse ends		Shape | Reverse Ends	
Bring to front	Ctrl+F	Shape | Bring to Front	
Send to back	Ctrl+B	Shape | Send to Back	
Center drawing		Tools | Center Drawing	

Resizing the Shape

Use the following procedure to resize (scale) the shape:

1. Click the shape.
2. Move the cursor over an inside corner handle.
3. Notice that the cursor changes to a diagonal double arrow.
4. Drag the corner handle away from the shape.
5. Notice the thin outline of the shape growing larger.
6. Release the mouse button. The shape is larger.

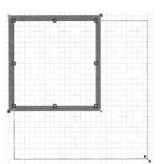

Stretching the Shape

Use the following procedure to stretch the shape:

1. Click the shape.
2. Move the cursor over an inside midpoint handle.
3. Notice that the cursor changes to a double arrow.
4. Drag the midpoint handle away from the shape.
5. Notice the thin outline of the shape stretching.

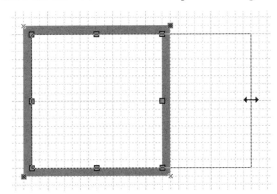

6. Release the mouse button. The shape is longer.

Changing the Wall Thickness

Use the following procedure to change the thickness of the walls:

1. Click the shape.
2. Move the cursor over an outside corner handle.

3. Notice that the cursor changes to a four-headed arrow and a small hint box explains the function of the handle ("Adjust Wall Thickness").

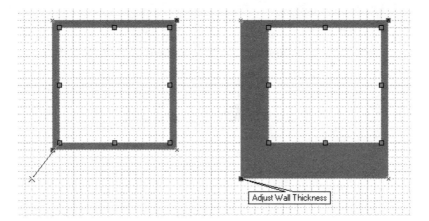

4. Drag the corner handle away from the shape.
5. Notice the thin line stretching out. The length of the line indicates the new width of the two walls connected to the handle. Move diagonally to change the width of both the horizontal and vertical walls.
6. Release the mouse button. The wall is thicker.
7. Repeat the steps above to change the thickness of the other two walls.

Rotating the Shape

Use the following procedure to rotate the shape by 90 degrees:

1. Click the shape.
2. Select **Shape | Rotate Left**.
3. Notice the shape has rotated counterclockwise by 90 degrees. The status line reports "Angle = 90 deg."

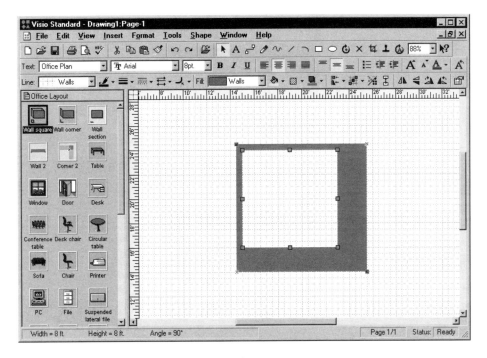

4. Repeat to rotate another 90 degrees.

Free Rotating the Shape

Use the following procedure to free rotate the shape:

1. Click the shape.
2. Press the **Rotation Tool** button on the toolbar.
3. Notice the shape has three new handles (green circles):
    ➤   **Center of Rotation**: the center handle is a green circle with a small cross; the shape rotates around this handle. This handle can be moved to change how the shape is rotated.
    ➤   **Rotation Handle**: the other two handles are green circles; use these handles to rotate the shape around the center of rotation.
4. Move the cursor over one of the rotation handles. The cursor changes to a pair of curved arrows.
5. Notice the thin outline of the shape rotating as you drag the cursor. The status line reports the angle.

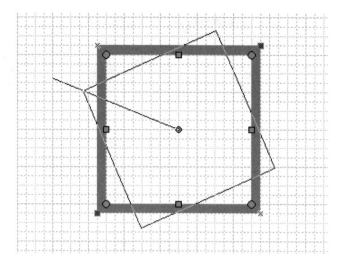

6. Press the Pointer Tool button on the toolbar to get out of rotation mode.

Flipping the Shape

Use the following procedure to flip the shape:

1. Click the shape.
2. Select **Shape | Flip Horizontal**.
3. Notice the shape has mirrored about its center point. The status line reports the new angle.

The Size and Position Dialog Box

Use the following procedure to precisely change the size and rotation angle of a shape by typing numbers:

1. Click the shape.
2. Select **Shape | Size & Position**.
3. Notice the Size & Position dialog box. Text has changed from black to blue when values previously change from the shape's original size and position.

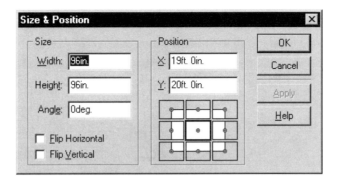

4. To change the size of the shape, type new values in the Width and Height boxes.

5. To change the rotation, type a new angle in the Angle box.

6. To change the center of rotation, type new distances in the X and Y boxes.

7. Alternatively, change the center of rotation to one of nine locations by clicking on the **Position** boxes.

8. To flip horizontally or vertically, select the **Flip Horizontal** or **Flip Vertical** check box.

9. To reverse ends, select both the **Flip Horizontal** and **Flip Vertical** check boxes.

10. Click **OK**.

Bring to Front and Send to Back

Use the following procedure to change the visibility of overlapping shapes:

1. Click the shape.

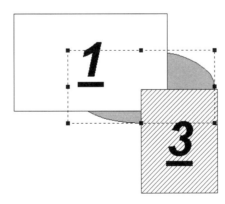

2. Select **Shape | Bring to Front**.

3. Notice how the shape moves on top of overlaying shapes.

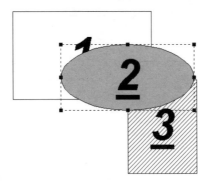

4. Select **Shape | Send to Back**.

5. Notice how the shape moves behind the overlaying shapes.

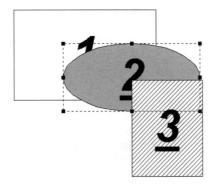

Center Drawing

Use the following procedure to center the drawing:

1. Select **Tools | Center Drawing**.

2. Notice that Visio centers the drawing on the page.

Hands-On Activity

In this activity, you use the size functions to resize the Wall square shape. Begin by starting Visio. Ensure Visio is running and is displaying the Office Layout stencil.

1. Drag the **Desk Chair** shape into the drawing. If necessary, zoom in to make the chair look larger.

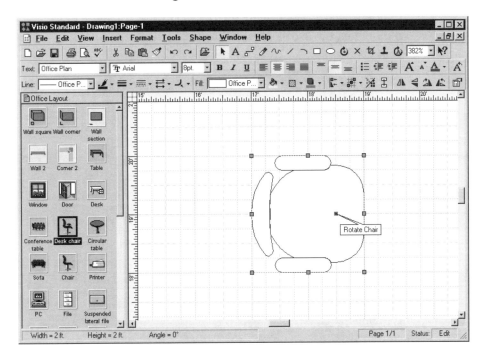

2. Click the **Desk Chair** shape. Notice the eight handles (green squares) and one rotation handle.

3. Select **Shape | Size & Position**. Notice the Size & Position dialog box.

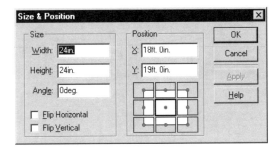

4. Type **45** in the Angle box.

5. Type **19** for both the X and Y position.

6. Select the lower left corner for Position.

7. Notice the Width and Height text changes from black to green. (The next time you access the dialog box, the color of the text will be blue.)

🖐 **Point of Interest:** The green text in a dialog box indicates the settings that have changed. The next time you open this dialog box, the text is blue, indicating the settings were changed earlier.

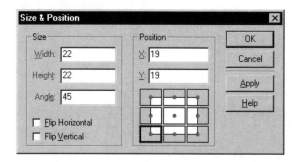

8. To see the effect of these changes without dismissing the dialog box, click the **Apply** button.

9. Click **OK**.

10. Notice Visio rotates the chair shape and moves the shape to a different position on the page.

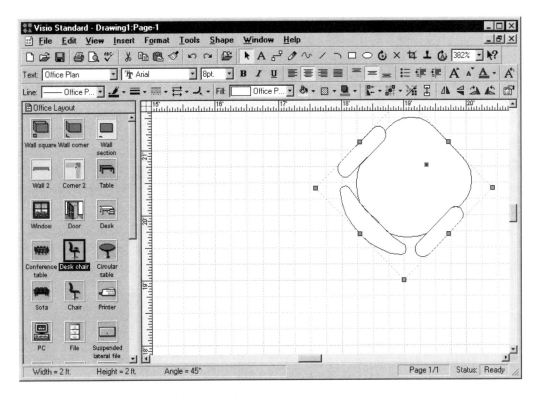

11. Exit Visio with **Alt+F4**.

This completes the hands-on activity for sizing and positioning shapes.

Module 10

Placing Multiple Shapes
Edit | Duplicate, Shape | Operations | Offset

Uses

Once a shape has been placed in the drawing, you make copies of the shape in the drawing. When you change a shape, you don't want to apply those changes to every additional shape; instead, you make copies (or *duplicates*) of the modified shape.

Similarly, sometimes you want to create shapes parallel to existing shapes. Visio allows you to *offset* simple lines and arcs.

Procedures

Before presenting the general procedures for placing multiple shapes, it is helpful to know about the shortcut keys. These are:

Function	Keys	Menu
Duplicate (copy)	Ctrl+drag or Ctrl+D	Edit | Duplicate
Repeat last action F4		Edit | Repeat
Offset		Shape | Operations | Offset

Making a Copy

Use the following procedure to copy a shape within the drawing:

1. Select a shape.

2. Hold down the **Ctrl** key. Notice that a small + (plus) sign appears next to the arrow cursor. Visio is reminding you that it will make a copy of the shape, rather than move it.

3. Drag the shape, and release the **Ctrl** key. Notice that an exact copy of the shape appears.

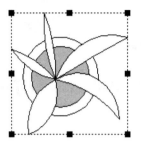 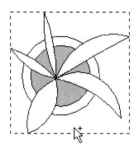

✍️ **Point of Interest:** To copy (or move) a shape in precisely horizontal and vertical direction, hold down the **Shift** key. This constrains movement of the cursor.

Duplicating More Than One Copy

Use the following procedure to duplicate a shape several times at the same distance apart:

1. Select shape.
2. Hold down the **Ctrl** key, drag the shape, and let go.
3. Notice that an exact copy of the shape appears.
4. Press **F4** to repeat the action.
5. Repeat pressing **F4** until you have enough copies.

6. If you make too many copies, press **Ctrl+Z** to undo the copy action.

Making an Offset Copy (Visio Technical Only)

Use the following procedure to offset lines and arcs:

1. Draw a line or arc.

2. Select the line (or arc).

3. Select **Shape | Operations | Offset**. Notice the Offset dialog box. If the shape is one that Visio cannot offset (such as a chair or plant shape), the Offset command is grayed out.

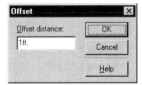

4. Type an offset distance in the **Offset Distance** text box.

5. Click **OK**. Notice that Visio places a copy of the shape on either side. If the shape is quite complex, Visio does a poor offset job, resulting in a simple straight line.

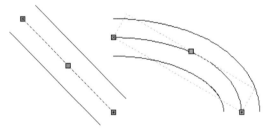

Hands-On Activity

In this activity, you use the duplicate function. Ensure Visio is running and is displaying the Office Layout stencil.

1. Drag the **Bookshelf** shape into the drawing.

2. Hold down both the **Ctrl** and **Shift** keys.

3. Drag a copy of the **Bookshelf** shape to the right so that it touches the first bookshelf.

4. Let go of the **Ctrl** and **Shift** keys.

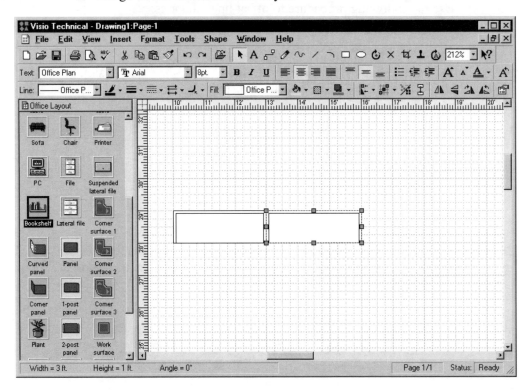

5. Press **F4** twice to place two more bookshelves.

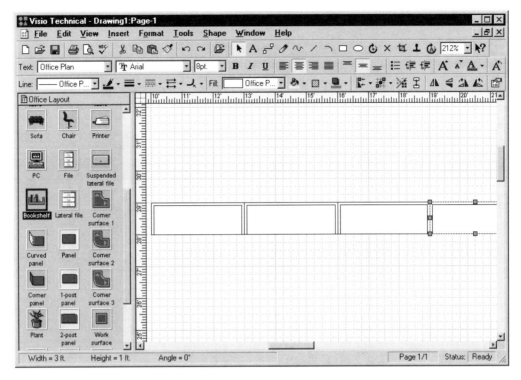

6. If necessary, use one of the **Zoom** commands to see all four bookshelves.

7. Press **Alt+F4** to exit Visio.

This completes the hands-on activity for placing multiple shapes.

Module 11

Connecting Shapes Together

Tools | Connect Shapes

Uses

The **Connector** tool draws connections (intelligent lines) between shapes. A *connection* is a line that attaches to specific points on a shape, called *connection points*. The connector stretches longer and shorter when the connected shape is moved.

The connection point on a shape is shown by a small blue x; when a connection is successful, the connection point becomes a red square.

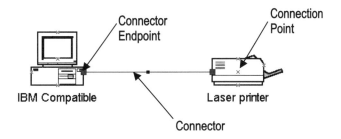

There are three ways to connect shapes:

1. Click the **Connector** tool button, then drag shapes from the stencil to the page. The shapes are automatically connected.

2. Drag shapes to the page. Then use the **Connector Tool** to manually connect the shapes.

3. Use the **Connect Shapes** tool, which automatically places a connection between selected shapes.

If the shape does not have a connection point (the small blue x) in a convenient location, you add a connection point to the shape with the **Connection Point** tool.

To move a connection (or connection point), select the connection and drag it to a new location. A selected connection becomes a green-black dashed line; a selected connection point becomes magenta (pink).

If the drawing is too cluttered with many connection points, select **View | Connection Points** to turn off the display of those little blue x's. To delete a connection (or connection point), select the connection and press the **Delete** key.

Procedures

Before presenting the general procedures for using the **Connector** tool, it is helpful to know about the buttons and shortcut keys. These are:

Function	Keys	Menu	Toolbar Icon
Connector Tool	Ctrl+3		
Connect Shapes	Ctrl+K	Tools | Connect Shapes	
Connection Point Tool			
Toggle display of connector tool		View | Connection Points	

Connecting Shapes Automatically During Dragging

Use the following procedure to automatically connect shapes as you drag them onto the page:

1. Press the **Connector** tool button. Notice that the cursor changes to a black arrow with a zig-zag connector near it.
2. Drag a shape from the stencil to the page.
3. Drag another shape from the stencil to the page.
4. Notice that Visio draws a connector between the first and second shape.
5. As you drag additional shapes to the page, Visio connects them in sequential order: the most recent shape is connected to the next most recent shape.

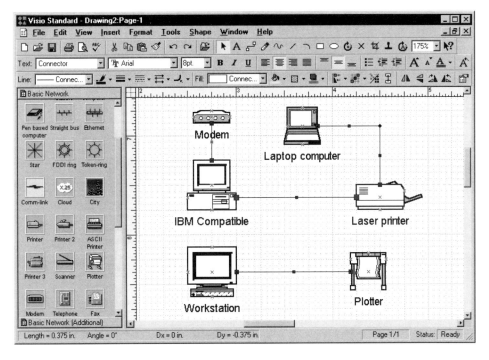

Connecting Existing Shapes Automatically

Use the following procedure to automatically connect shapes that are already on the page:

1. Press **Ctrl+A** to select everything on the page.

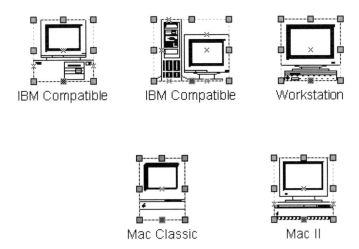

2. Select **Tools | Connect Shapes**. Notice that Visio draws a connector between the shapes approximately in the order that you placed them in the drawing.

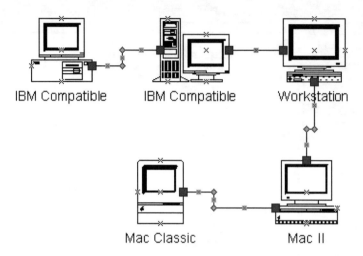

3. You may need to move some of the shapes to straighten out the connectors and make them look more pleasing. Notice how the connectors "stick" to the shapes as you move them around. This is one of Visio's most powerful features.

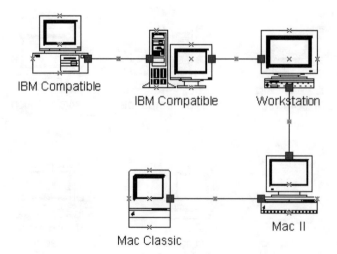

Point of Interest: Visio connects shapes automatically in the order you placed the shapes. If you plan to use the automatic connection feature, think ahead about shape placement, otherwise you may end up with the complicated connections shown in the figure below.

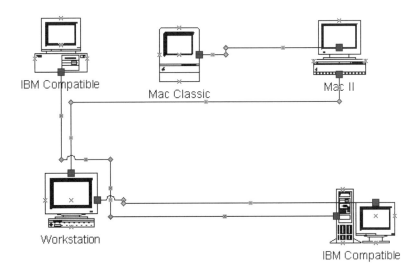

Connecting Two Shapes Automatically

Use the following procedure to automatically connect two shapes already on the page:

1. Select a shape.
2. Press the **Shift** key and select another shape.
3. Select **Tools | Connect Shapes**.
4. Notice that Visio draws a connector between the first and second shape.

Connecting Shapes Manually

Use the following procedure to add connections between shapes:

1. Press the **Connector** tool button.
2. Click on a shape's connection point (the small blue x).
3. Drag the connector to another shape's connection point.

4. Notice that the connector's endpoints are red squares; this tells you the end-point successfully connected to a connection point. If the square is green, the endpoint is not connected to a connection point.

Adding a Connection Point

Use the following procedure to add a connection point to a shape:

1. Press the **Connection Point** tool button on the toolbar.
2. Move the cursor to the location where you want the connection point.
3. Press the **Ctrl** key, then click.
4. Notice the magenta (pink) x; this is the new connection point.

Hands-On Activity

In this activity, you use the functions of the connection tools. Begin by starting Visio. Then open a new document using the Network template supplied with Visio.

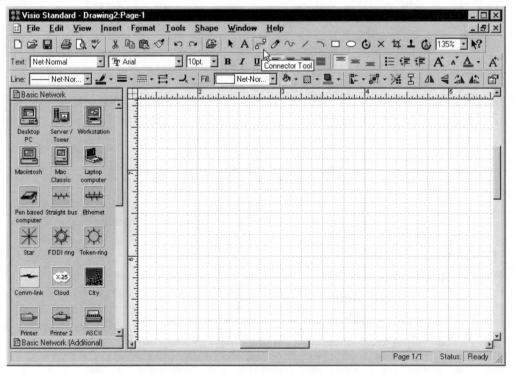

1. Press the **Connector** tool ⬜ button.

2. Drag the **Desktop PC** shape from the Network stencil to the upper left area of the page.

3. Drag the **Server/Tower** shape from the stencil to the center area of the page.

4. Notice that Visio draws a connector between the first and second shape.

5. Drag the **Workstation** shape to the lower right area of the page.

6. Notice that Visio draws a second connector between the second and third shape.

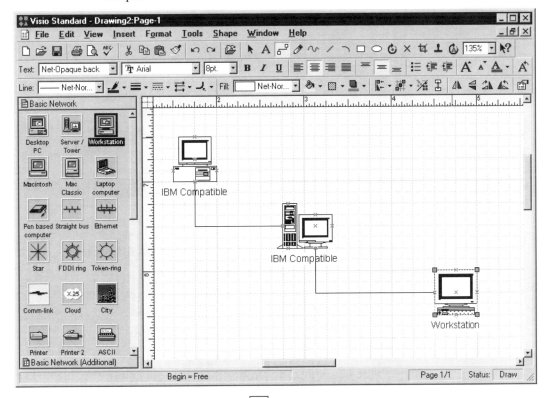

7. Press the **Pointer** tool ⬜ button on the toolbar.

8. Drag the **Macintosh** shape to the lower left area of the page.

9. Drag the **Mac Classic** shape to the upper right area of the page.

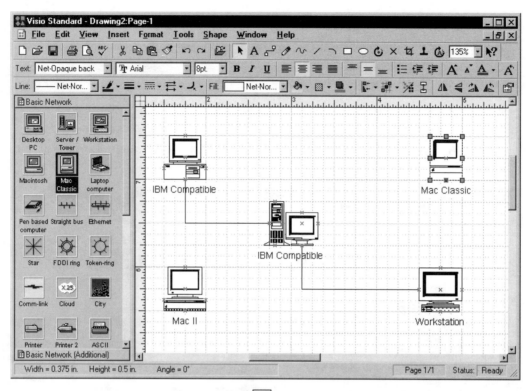

10. Press the **Connector** tool 🔲 button on the toolbar.

11. Click a connection point (small blue x) of the Macintosh shape.

12. Hold down the mouse button.

13. Drag the connector to a connection point on the Server/Tower shape.

14. Release the mouse button. Notice that Visio connects the two shapes and that a pair of red squares appear at the connection points.

15. Press the **Pointer** tool button.

16. Select the **Mac Classic** shape. Notice that Visio surrounds the selected shape with green squares, which acknowledge the selection.

17. Hold the **Shift** key and select the **Server/Tower** shape. Notice that Visio surrounds the selected shape with cyan (light blue) squares, which acknowledges adding to the selection.

18. Select **Tools | Connect Shapes**.

19. Notice that Visio automatically connects the two shapes.

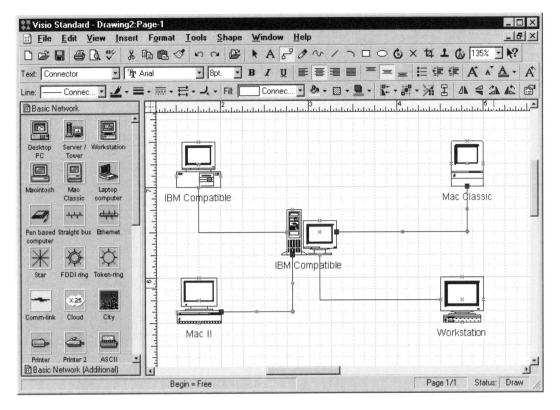

20. Press **Alt+F4** to exit Visio.

This completes the hands-on activity for connecting shapes.

Module 12

Cutting, Copying, and Pasting

Edit | Cut, Copy, Paste, Paste Special

Uses

The **Cut**, **Copy**, and **Paste** selections of the **Edit** menu are used to move shapes and text from one location to another. With all software that runs under Windows, nearly any object can be copied and pasted between applications. For example, an Excel spreadsheet can be pasted in a Visio drawing. A Visio drawing can be pasted in a Word document.

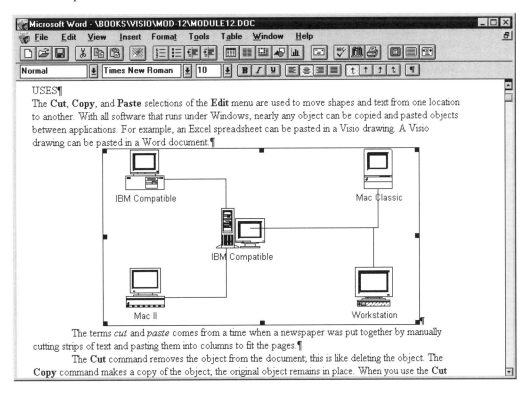

The terms *cut* and *paste* comes from a time when a newspaper was put together by manually cutting strips of text and pasting them into columns to fit the pages. Today, with computers, it is more common to copy, rather than cut.

The **Cut** command removes the object from the document; this is like deleting the object. The **Copy** command makes a copy of the object; the original object remains in place. Do not confuse the **Copy** command with the **Duplicate** command. **Copy** can be used for duplication; **Duplicate** does not send a copy to the Clipboard.

When you use the **Cut** and **Copy** commands, Windows moves the object to the Clipboard, which is a temporary holding area. The next time you use the **Cut** or **Copy** command, the new object replaces the previous one being stored in the Clipboard.

The **Paste** command copies the object from the Clipboard and places it in the document. You can use the **Paste** command several times in a row to paste the same object several times into one or more documents.

When an object is sent to the Clipboard, it is often sent in many different formats. To control the format that gets pasted in your document, use the **Paste Special** command, which displays a dialog box that lets you select a format. In contrast, the **Paste** command pastes the first format listed in the **Paste Special** dialog box.

Paste Special sometimes lets you link the object back to its source. This makes it easier to update the object. More about linking in Module 28.

The **Paste as Hyperlink** is a special command that only works when a hyperlink is in the Clipboard. This command pastes the hyperlink as a hyperlink in the Visio document. More about hyperlinks in Module 36.

Procedures

Before presenting the general procedures for cutting, copying, and pasting text and graphics objects, it is helpful to know about the shortcut keys. These are:

Function	Keys	Menu	Toolbar Icon
Cut	Ctrl+X	Edit | Cut	
Copy	Ctrl+C	Edit | Copy	
Paste	Ctrl+V	Edit | Paste	
Paste Special		Edit | Paste Special	
Paste Hyperlink		Edit | Paste as Hyperlink	

Cutting Text and Graphic Objects

Use the following procedure to cut an object to the Clipboard:

1. Select a shape by clicking on it so that green square handles appear on a green, dotted frame.

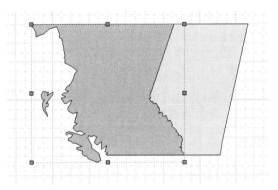

2. Select **Edit | Cut** (or press **Ctrl+X**) to remove the selected shape from the page and place it in the Clipboard

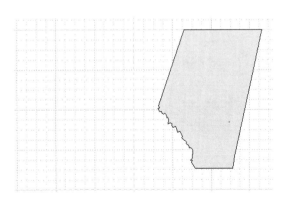

119

Copying Text and Graphic Objects

Use the following procedure to copy an object to the Clipboard:

1. Select a shape by clicking on it.
2. Select **Edit | Copy** (or press **Ctrl+C**) to copy the selected shape and place it in the Clipboard.

Pasting Text and Graphic Objects

Use the following procedure to paste an object into the page:

1. Select **Edit | Paste** (or press **Ctrl+V**) to paste whatever is in the Clipboard onto the page.

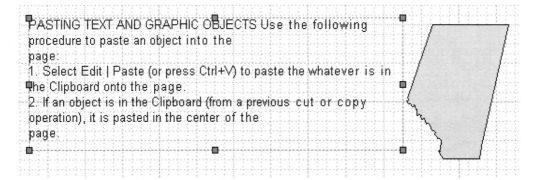

2. If an object is in the Clipboard (from a previous cut or copy operation), it is pasted in the center of the page.
3. Drag the object into place and resize if necessary.

Paste Special

Use the following procedure to control the paste format of an object:

1. Select **Edit | Paste Special**. Notice that Visio displays the Paste Special dialog box.

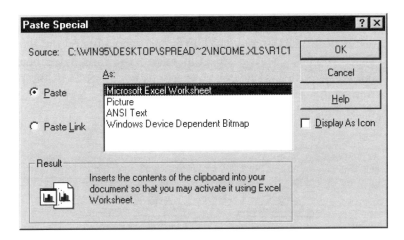

The Paste Special dialog box contains the following options:

➤ **Paste** pastes the object without linking.

➤ **Paste Link** pastes the object with a link back to the source application. If the source application does not allow linking, the **Paste Link** radio button is grayed out.

➤ **As** describes the formats available in the Clipboard.

➤ **Display As Icon** pastes the object as an icon. Selecting this option displays a default icon, along with the **Change Icon** button to let you select another icon.

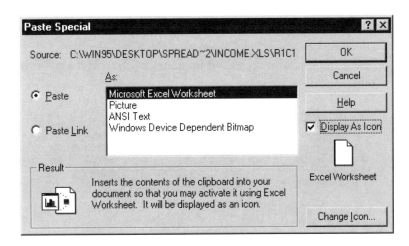

3. Select a format from the **As** list box. When an object is available in several
   different formats, it may look different, depending on the format you select.
   When you paste a Visio object in a non-Visio format, the object loses all its
   intelligence, such as layers, connections points, and custom properties. The
   figure below shows a Visio shape pasted as a Visio object (left) and as a bitmap, right:

Satellite dish   Satellite dish

In particular, text is greatly affected by paste formats. For example, the following Excel spreadsheet pasted in Picture format retains much of the
formatting:

Received Income 1995-6			
Date Invoiced	**Project**	**Acct No**	**Cheque**
5 Sep 95	CADinf - 2 articles	4107	¼□_□S□
27 Jun 95	CADinf - 1 article	4107	¼□\□S□
15 Sep 95	Typesetting-T2 QR	4510	–□À□V□[□
	Book order	4810	–□…□S

The same spreadsheet pasted in plain Text format loses all formatting, except
tabs:

Received Income 1995-6				
Date Invoiced	Client	Project	Acct No	Cheque
5 Sep 95		CADinf - 2 articles	4107	$160.00
27 Jun 95		CADinf - 1 article	4107	$100.00
15 Sep 95		Typesetting-T2 QR	4510	$3,860.00
		Book order	4810	$35.00

The spreadsheet pasted in Bitmap format retains the look of a spreadsheet,
complete with column and row headers:

	A	C	D	E
1	**Received Income 1995-6**			
2	Date Invoiced	Project	Acct No	Cheque
3	5 Sep 95	CADinf - 2 articles	4107	$160.00
4	27 Jun 95	CADinf - 1 article	4107	$100.00
5	15 Sep 95	Typesetting-T2 QR	4510	$3,860.00
6		Book order	4810	$35.00

Here is the spreadsheet pasted as an icon:

Income.xls

4. Click **OK**. Notice the object is pasted in the center of the page.
5. Drag the object into place and resize if necessary.

Hands-On Activity

In this activity, you use the cut, copy, and paste functions. Begin by starting Visio. Then open the **Basic Network** template file.

1. Click the **Basic Network (Additional)** stencil title bar to make it visible.

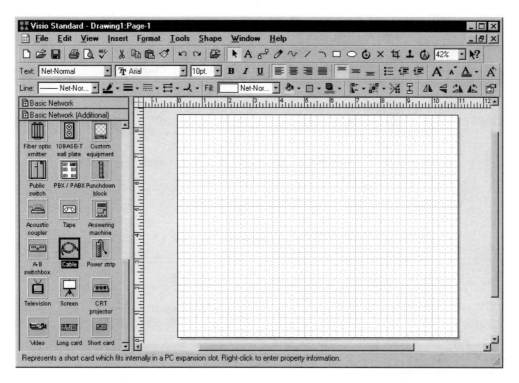

2. Drag the **Cable** shape into the drawing. If necessary, zoom to get a better view.

3. Ensure the shape is selected (green handle squares surround it), then press **Ctrl+C** to copy the shape to the Clipboard. Or, click the **Copy** icon 📋 on the toolbar. Or, right-click and select **Copy** from the menu.

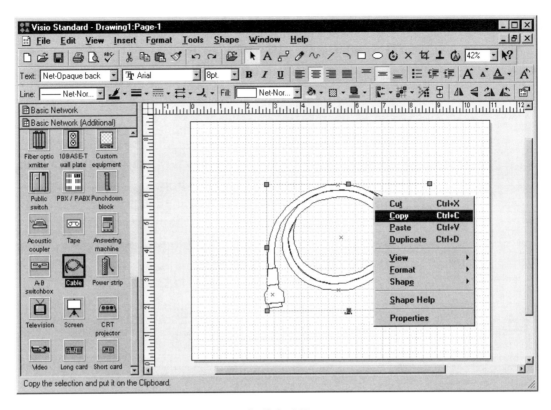

4. Select **Insert | Page** and click **OK** to create a new page.

5. Press **Ctrl+V** to paste the copied shape. Notice that Visio places the shape in the center of the page.

6. Notice that the shape retains its intelligence. For example, the connection points (small blue x) are present.

7. Select **Edit | Paste Special**. Notice the Paste Special dialog box with several format options:

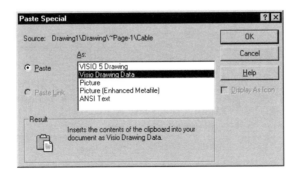

> ➤ **Visio 5 Drawing:** The drawing in Visio 5 format.

> ➤ **Visio Drawing Data:** The drawing in a format read by earlier versions of Visio.

> ➤ **Picture:** A vector format known as Windows Metafile or WMF, for short.

> ➤ **Picture (Enhanced Metafile):** A newer version of WMF found in Windows 95.

> ➤ **ANSI Text:** The text portion of the shape.

8. Ensure that **Picture** is the selected format and click **OK**. Notice that Visio places the graphic object in the center of the page.

9. Move the two pasted cable objects apart. Notice that the cable pasted as a Picture lacks the connection point.

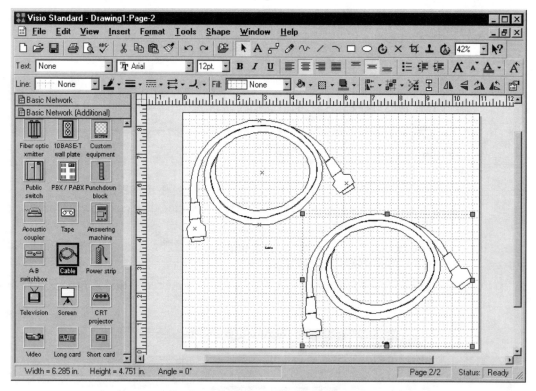

10. Select **Edit | Paste Special**.

11. Select **ANSI Text** and click **OK**. Notice that Visio pastes the word "Cable" in the center of the drawing.

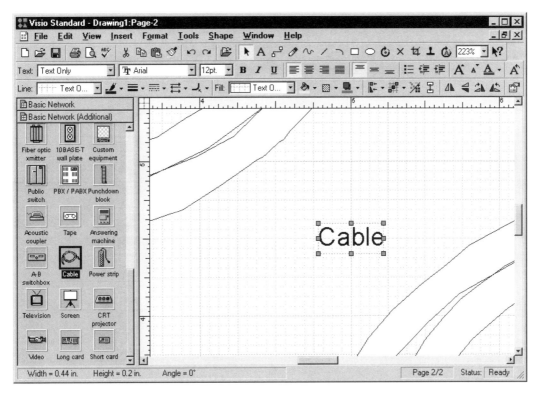

12. Press **Alt+F4** to exit Visio.

This completes the hands-on activity for cut, copy, and paste. These functions are identical in all Windows applications. Only the available formats in the Paste Special dialog box differ.

Module 13

Formatting Shapes

Format | Line, Corners, Fill, Shadow

Uses

The formatting functions, found on the **Format** menu, are used to change the look of shapes. You can change the look of the lines that make up all shapes. "Line" is a generic term applying to not just lines but also circles, arcs, curves, and rectangles and ellipses. The **Format | Line** command lets you select from:

➤ 24 patterns of dashes and dots.

➤ Weight ranging from 1 to 17 pixels in width, plus custom weights.

➤ 12 colors, 12 shades of gray ranging from black to white, plus custom colors selected from the entire 16.7 million Windows palette.

➤ End caps with round or square ends to the line.

➤ 46 different line ends, such as arrowheads, in seven sizes ranging from Very Small to Colossal.

When two lines (or arcs or curves) meet, the **Format | Corners** command lets you select from seven different radii of rounded corners, as well as the default (no rounding—a sharp corner), and a custom radius.

Areas are formed by lines, circles, arcs, curves, and rectangles and ellipses. Usually, these are empty or are filled with white color. The **Format | Fill** command lets you select from 24 patterns of lines and dots, and 24 colors and shades of gray, and 16 gradient fills for the foreground and background. "Foreground" refers to the lines and dots that make up the pattern, while "background" refers to the underlying area.

Every object—whether shape, text, or a pasted object—can cast a shadow in a Visio drawing By default, no shadow is cast; adding the shadow helps make the object stand out on the page from other objects. The **Format | Shadow** command lets you select a pattern and color for the shadow. Be careful, though, some printers print the shadow as black, no matter how you format it.

Procedures

Before presenting the general procedures for formatting, it is helpful to know about the shortcut keys. These are:

Function	Keys	Menu	Toolbar Icon
Line Color		Format | Line	
Line Ends		Format | Line	
Line Pattern		Format | Line	
Line Weight		Format | Line	
Format Corners		Format | Corners	
Fill Pattern	F3	Format | Fill	
Fill Color		Format | Fill	
Format Shadow		Format | Shadow	

Notice that the **Format | Line** menu selection also displays **Corner** options. In the same way, the **Format | Fill** selection displays **Shadow** options.

Formatting Lines

Use the following procedure to format the look of lines, circles, arcs, curves, rectangles, and ellipses:

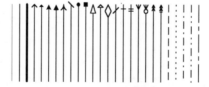

1. Select the shape.
2. Select **Format | Line**. Notice that Visio displays the Line dialog box.
3. To change the line pattern, click on the **Pattern** list box and select one of the 24 patterns. When you select pattern **None**, the line becomes invisible.

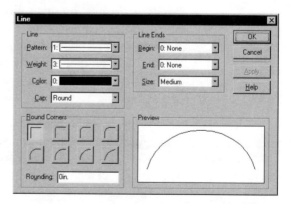

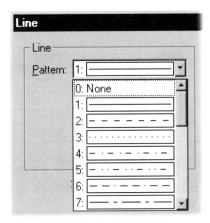

4. Notice that as you make selections from this dialog box, the **Preview** window displays the effect of the change on an arc.

5. To see the effect of the changes without exiting the dialog box, click **Apply**. You may need to move the dialog box aside to see the affected shapes.

6. Click **OK** to exit the Line dialog box.

Change the Weight of Lines

Use the following procedure to change the width (weight) of lines:

1. Select the shape.

2. Select **Format | Line**. Notice that Visio displays the Line dialog box.

3. Click the **Weight** list box and select one of the widths ranging from 1 to 17:

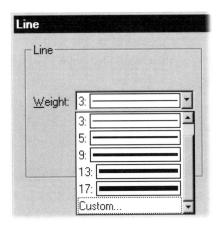

4. Notice that the weight numbers correspond to the following widths:

Weight	Points	Inches	Millimeters
1	0.24pt	0.0033"	0.085mm
3	0.72pt	0.0100"	0.254mm
13	3.12pt	0.4333"	1.10mm
17	4.08pt	0.5667"	1.44mm

5. To choose a line weight other than the 17 preprogrammed weights, select
   Custom from the Weight list box. Notice that Visio displays the Custom Line
   Weight dialog box.

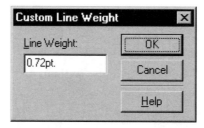

6. Type a number and follow it with a unit, such as **.1in**. Visio allows the following units:

Unit	Meaning
"	inch
in	inch
'	foot
ft	foot
mi	miles
mm	millimeter; 25.4mm = 1"
cm	centimeter
m	meter
pt	point; 72.727272 points = 1"
p	pica; 6 picas = 1"

7. Click **OK** to dismiss the Custom Line Weight dialog box.
8. Click **OK** again to dismiss the Line dialog box.

Changing the Line Color

Use the following procedure to change the color of selected lines:

1. Select the shape.
2. Select **Format | Line**. Notice that Visio displays the Line dialog box.
3. Click on the **Color** list box and select one of the 24 colors.

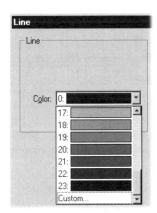

4. If you prefer a color not shown, select **Custom**. Notice that Visio displays the Color dialog box.

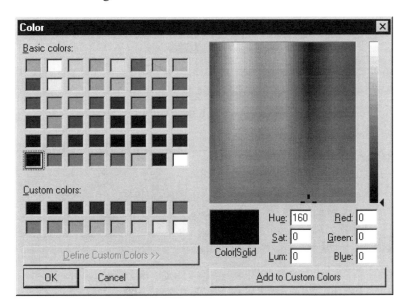

5. Click on one of the 16.7 million colors displayed.

6. Click **OK** to dismiss the Color dialog box.

7. Click **OK** to dismiss the Line dialog box.

Change the Line End Cap

Use the following procedure to change the end cap of lines:

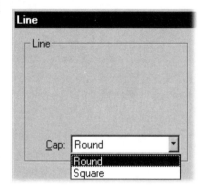

1. Select the line.

2. Select **Format | Line**. Notice that Visio displays the Line dialog box.

3. Click the **Cap** list box and select **Round** or **Square**.

4. Notice the difference between round and square end caps in the illustration below. In most cases, the difference between rounded and square corners is unnoticeable, unless applied to very wide lines or seen at a very high zoom level.

Selecting an Arrowhead

Use the following procedure to change the line ends:

1. Select the line.

2. Select **Format | Line**. Notice that Visio displays the Line dialog box.

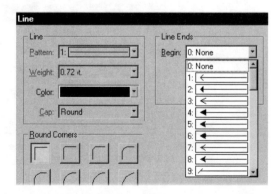

3. Click the **Begin** list box to select the beginning of the line.

4. Click the **End** list box to select the ending of the line.

5. Click the **Size** list box to select the size of the line end.

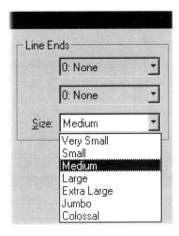

6. Notice in the illustration below the size of the arrowhead changing from very small (left) to colossal (right).

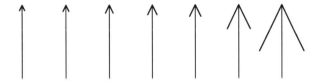

7. Click **Apply** to make the change without dismissing the dialog box.
8. Click **OK** to make the changes and dismiss the dialog box.

Rounding Corners

Use the following procedure to round the corners of rectangles, or two or more lines, arcs, and curves:

1. Select the shape (or rectangle or ellipse); or, select two or more lines, arcs, or curves.
2. Select **Format | Corners**.
3. Notice that Visio displays the Corners dialog box.
4. To change the corner, click on one of the round corner buttons.
5. Or, type a value for the rounded corner radius in the **Rounding** text box.

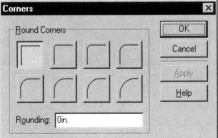

6. Click **Apply** to make the change without dismissing the dialog box.

7. Click **OK** to make the changes and dismiss the dialog box.

Applying a Fill Color and Pattern

Use the following procedure to change the fill of rectangles, circles and ellipses, or the area created by lines, arcs, and curves:

1. Select the shape or lines, arcs, or curves.

2. Select **Format | Fill**.

3. Notice that Visio displays the **Fill** dialog box.

4. To change the pattern, click on the **Pattern** list box and select one of the 41 patterns.

5. Notice that pattern #0 is **None**, which means the object is transparent.

6. To change the color of the pattern lines and dots, click on the **Foreground** list box and select one of the 24 colors. Or, select **Custom** to choose a color from the 16.7 million available in Windows.

7. To change the color underneath the pattern of lines and dots, click on the **Background** list box and select one of the 24 colors.

8. Notice that Visio displays the changes you make in the **Preview** box. The illustration below show no fill (#0), solid fill (#1) and several variations of gradient fill. Notice the difference between no fill and white solid fill: The grid lines show through the no-fill rectangle.

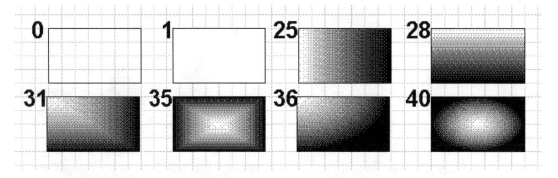

9. Click **Apply** to make the change without dismissing the dialog box.

10. Click **OK** to make the changes and dismiss the dialog box.

Creating a Drop Shadows

Use the following procedure to change the shadow:

1. Select the shape or lines, arcs, or curves.
2. Select **Format | Shadow**.
3. Notice that Visio displays the Shadow dialog box.

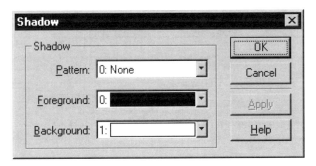

4. To change the pattern, click on the **Pattern** list box and select one of the 25 patterns.
5. To change the color of the pattern lines and dots, click on the **Foreground** list box and select one of the 24 colors. Or, select **Custom** to choose a color from the 16.7 million available in Windows.
6. To change the color underneath the pattern of lines and dots, click on the **Background** list box and select one of the 24 colors.
7. Click **Apply** to make the change without dismissing the dialog box. The following illustration shows some samples of the shadow patterns. Notice that if a shape is not filled, the shadow "shows through." When a shadow is applied to a shape, such as the laser printer, all lines making up the shape cast a shadow. You cannot select a shape by clicking on its shadow.

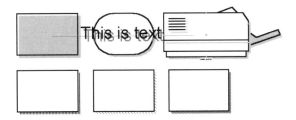

8. Click **OK** to make the changes and dismiss the dialog box.

Quick Formatting

To quickly format a shape, Visio provides a number of buttons that cycle through the most common formats:

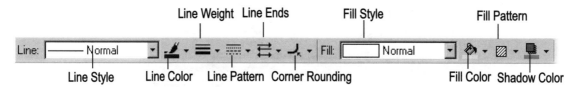

1. Select one or more shapes.
2. Click the arrow next to the **Line Color** ![button] button. Notice that Visio displays a small palette of colors.
3. Select a color.
4. Alternatively, select **No Line** to make the object invisible, or select **More Line Colors** to display the **Color** dialog box.
5. A similar procedure works for changing the line weight, patterns, ends, corner rounding, fill color, fill pattern, and shadow color.

Hands-On Activity

In this activity, you use the shape formatting functions. Begin by starting Visio. Then open **Charts and Graphs.Vst** template file found in the **Business Diagram** folder.

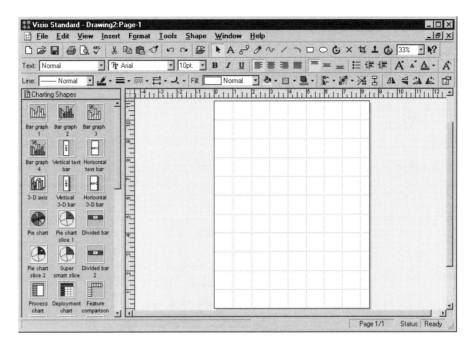

1. Drag the **3-D axis** shape into the drawing.
2. Adjust the zoom level to clearly see the shape.

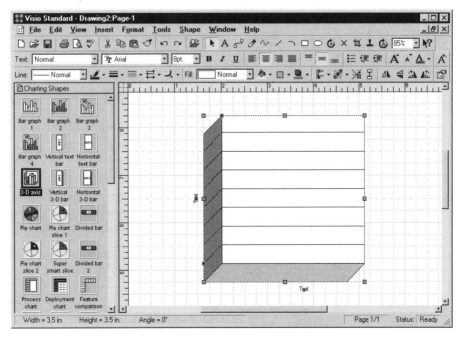

3. Click the arrow next to the **Line Color** button. Notice the color palette.

4. Click the blue color square. Notice that the shape's lines turn blue.

5. Click the arrow next to the **Line Weight** button. Notice the palette of line weights.

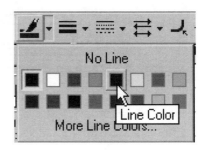

6. Select the widest line weight. Notice the shape's line grow thicker.

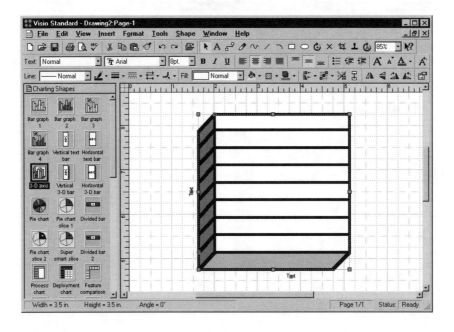

7. Click the arrow next to the **Line Pattern** button. Notice the palette of patterns.

8. Select the first dashed pattern. Notice that the lines change to a dashed pattern.

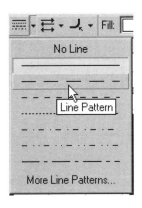

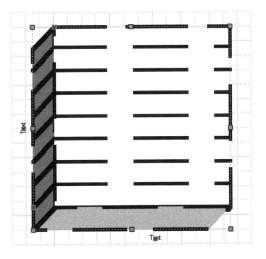

9. Click the arrow next to the **Line Ends** button. Notice the palette of arrowheads.

10. Select the first double-ended pattern. Notice that the lines now have arrowheads.

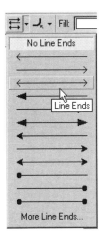

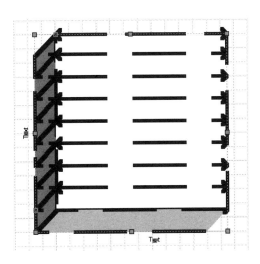

11. Click the arrow next to the **Corner Rounding** button. Notice the palette of corners.

12. Select the last corner pattern. Notice that the two rectangles now have round ends.

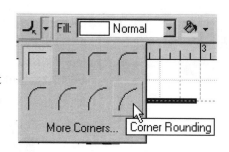

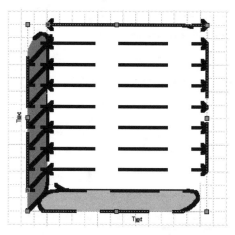

13. Click the arrow next to the **Fill Color** button. Notice the palette of colors.

14. Select the yellow color square. Notice that the shape fills with yellow color.

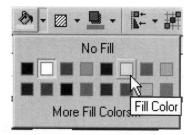

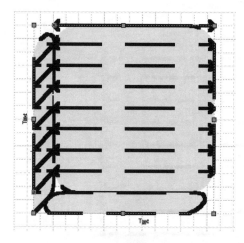

15. Click the arrow next to the **Fill Pattern** button. Notice the palette of patterns.

16. Select the diagonal line pattern. Notice that the shape's yellow fill takes on the diagonal striping.

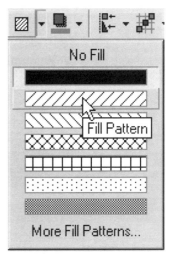

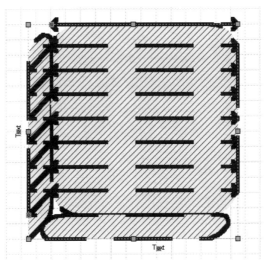

17. Click the arrow next to the **Shadow Color** button. Notice the palette of colors.

18. Select the first shade of gray. Notice that the shape casts a shadow.

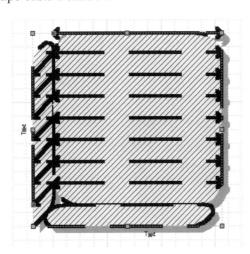

19. Press **Alt+F4** to exit Visio. Click **No** in response to the Save Changes dialog box.

This completes the hands-on activity for formatting shapes in the drawing.

Module 14

Formatting Text

Format | Text, Shape | Edit Text

Uses

The **Text** formatting function, found on the **Format** menu, changes the look of text in a Visio drawing. The command displays a tabbed dialog box that lets you change the font, size, case, position, language, style, horizontal and vertical alignment, margins, text and background color, tab spacing, and bullet style. As you can see from that list, there are many ways to format text!

The **Edit Text** function, found on the **Shape** menu, selects text within a shape.

A *text block* can be a single word or many paragraphs. It is all of the text within a single alignment box. An *alignment box* is the rectangle that outlines all shapes, after you click the shape. The illustration below shows two text blocks:

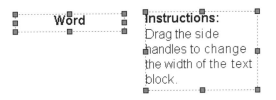

 Point of Interest: Before you can format text, you must select it. If you select only a portion of a block of text, the changes apply to the selected portion. This allows you to apply a different look to different parts of the text.

Procedures

Before presenting the general procedures for formatting text, it is helpful to know about the shortcut keys. These are:

Function	Keys	Menu	Toolbar Icon
Edit Text	F2	Shape | Edit Text	
Format Font	F11	Format | Text | Text	
Select Font Name			Tr Arial
Font Size			10pt.
Increase Font Size			A
Decrease Font Size			A
Bold			B
Italic			I
Underline			U
Font Color			A
Format Paragraph	Shift+F11	Format | Text | Paragraph	
Left Justify			
Center Justify			
Right Justify			
Full Justification			
Increase Indentation			
Decrease Indentation			
Increase Paragraph Spacing			
Decrease Paragraph Spacing			
Format Text Block		Format | Text | Text Block	
Top Alignment			
Middle Alignment			
Bottom Alignment			
Set Tabs	Ctrl+F11	Format | Text | Tabs	
Select Bullet		Format | Text | Bullet	

Selecting All Text In a Shape

Before you can format text, you must select it. Selected text reverses its color.

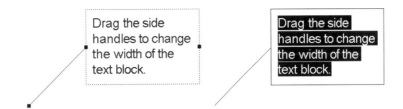

Use the following procedure to select text:

1. Click the shape containing text.
2. Press function key **F2**. Notice that Visio highlights the text by displaying it in reversed color, such as white text on black background.
3. Alternatively, select **Shape | Edit Text**.
4. Alternatively, double-click on the text.
5. Alternatively, right-click the shape and select **Format | Text**; all text in the shape is selected for formatting.

Selecting a Portion of the Text

To format a part of a text block (such as a single word), use the following procedure:

1. Select the shape.
2. Double-click the text. Notice that Visio highlights all of the text.
3. Drag the cursor over the characters you want to select.

The illustration shows a partial selection (left) and partially formatted text (right):

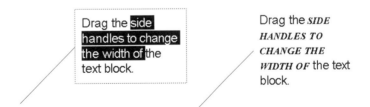

Selecting the Font

Use the following procedure to change the look of the text font:

1. Select the text.
2. Select **Format | Text**. Notice the Text dialog box.

147

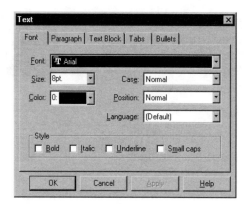

3. Click the **Font** list box to change the font (default = Arial). Notice that the list displays the names of fonts installed on your computer. A "TT" next to a font name indicates it is a TrueType font. A printer symbol means the font is installed on the printer.

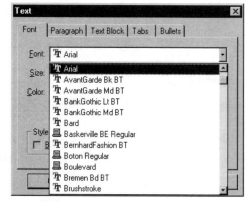

4. Select a font name. If necessary, scroll down the list. Click to select a font name.

5. Click **Apply** to see the font change without leaving the dialog box.

6. Click **OK** to dismiss the dialog box.

Changing the Size of the Text

Use the following procedure to change the size of the text:

1. Select the text.

2. Select **Format | Text**. Notice the Text dialog box.

3. Click the **Size** list box to change the height of the text (default = 8pt). The size is measured in *pt* (short for "point"), where one point equals 1/72" or 72 points = 1 inch.

4. Select a font size. For half-inch tall text, select 36pt.

5. Click **Apply** to see the font change without leaving the dialog box.

6. Click **OK** to dismiss the dialog box.

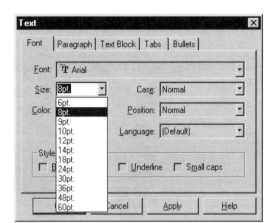

Changing the Color of the Text

Use the following procedure to change the color of the text:

1. Select the text.

2. Select **Format** | **Text**. Notice the Text dialog box.

3. Click the **Color** list box to change the color of the text (default = black). To change the background of the text, go to the **Text Block** tab of this dialog box.

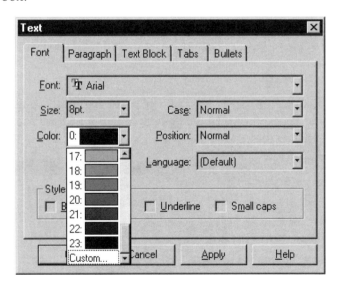

4. Select a color. You select from 12 colors and 12 shades of gray.

5. Alternatively, select **Custom** to display the **Color** dialog box.

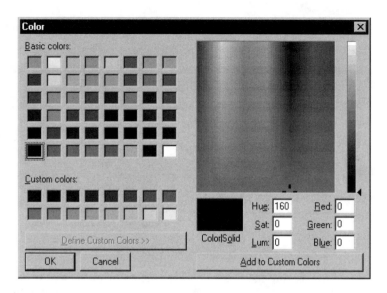

6. Select a color from the Windows palette of 16.7 million colors.
7. Click **OK** to dismiss the Color dialog box.
8. Click **Apply** to see the font change without leaving the dialog box.
9. Click **OK** to dismiss the dialog box.

Changing the Case of the Text

The *case* turns the text to all UPPERCASE or Initial Capitals (the first letter of every word is capitalized), which is excellent for titles. Use the following procedure to change the case of the text:

1. Select the text.
2. Select **Format | Text**. Notice the Text dialog box.
3. Click the **Case** list box.
4. Select **Normal**, **All Caps**, or **Initial Caps**.
5. Click **Apply** to see the font change without leaving the dialog box.
6. Click **OK** to dismiss the dialog box.

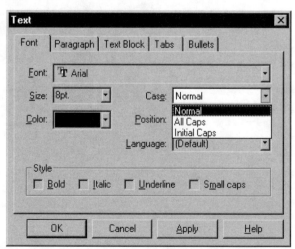

Changing the Position of the Text

The *position* moves the selected text higher (superscript) or lower (subscript), which is often used for footnotes and formulae. Use the following procedure to change the position of the text:

1. Select the text.

2. Select **Format | Text**. Notice the Text dialog box.

3. Click the **Position** list box.

4. Select **Normal**, **Superscript**, or **Subscript**.

5. Click **Apply** to see the font change without leaving the dialog box.

6. Click **OK** to dismiss the dialog box.

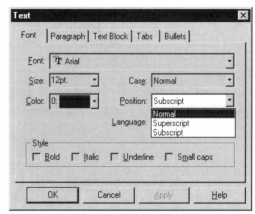

Changing the Language For Spell Checking

Use the following procedure to select the language:

1. Select the text.

2. Select **Format | Text**. Notice the Text dialog box.

3. Click on the **Language** list box.

4. Select a language name. This selection has no effect on the text until you use Visio's dictionary to check the spelling.

5. Click **OK** to dismiss the dialog box.

Changing the Style of the Text

Use the following procedure to change the look of the text:

1. Select the text.

2. Select **Format | Text**. Notice the Text dialog box.

3. Click one of the Style check boxes to change the look to any combination of the following: **bold**, *italic*, underline, or SMALL CAPS.

Normal text.
Boldface text.
Italicized text.
Underlined text.
SMALL CAPS TEXT.
Combination text.

4. Click **Apply** to see the format changes without leaving the dialog box.

5. Click **OK** to dismiss the dialog box.

Changing the Justification of a Paragraph

Use the following procedure to change the paragraph justification of the text:

1. Select the text.

2. Select **Format | Text**.

3. Select the **Paragraph** tab, which controls the formatting of paragraphs of text.

4. Click the **Horizontal Alignment** list box to change the justification, which is sometimes called "paragraph alignment" in other software.

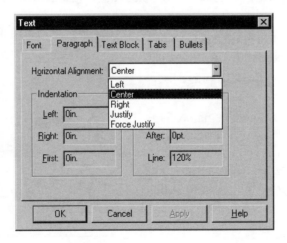

5. Select left, center (the default), right, justify, and force justify. *Force justify* means that the last line of text in the paragraph is forced to fit the width of the paragraph.

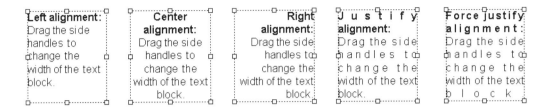

6. Click **Apply** to see the format changes without leaving the dialog box.

7. Click **OK** to dismiss the dialog box.

Changing the Indentation of a Paragraph

Use the following procedure to change the left, right, and first line indentation of a paragraph:

1. Select the text.

2. Select **Format | Text**.

3. Click the **Paragraph** tab, which controls the formatting of paragraphs. Notice the **Indentation** section has three text entry boxes. *Indentation* specifies how far in the text starts (left) and ends (right) from the margin.

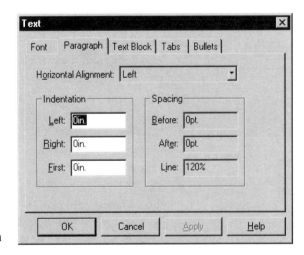

► The **Left** indentation "pushes" all text rightward from the left margin.

► The **Right** indentation pushes all text leftward from the right margin.

► The **First** indentation indents the first line. In most cases, you would use the **First** indentation if you want to create the look of a traditional paragraph with the indented first line. To create a *hanging indent*, type a negative number for the **First** indentation.

4. Click **Apply** to see the format changes without leaving the dialog box.

5. Click **OK** to dismiss the dialog box.

Changing the Spacing Within and Between Paragraphs

Use the following procedure to set the spacing between paragraphs, or change the spacing between text lines:

1. Select the text.

2. Select **Format | Text**.

3. Click the **Paragraph** tab, which controls the formatting of paragraphs. Notice the **Spacing** section has three text entry boxes. *Spacing* specifies the distance between lines of text.

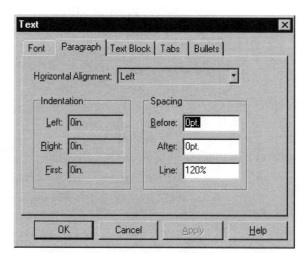

➤ The **Before** spacing increases the distance between the top of the paragraph and the preceding paragraph.

➤ The **After** spacing increases the distance between the bottom of the paragraph and the following paragraph.

➤ The **Line** spacing sets the distance between lines of text within a paragraph, measured as a percentage of the font size; default = 120%. For example, if the font size if 8pt, the line spacing is 8pt * 120% = 9.6pt, measured from the baseline of one line of text to the baseline of the following line of text.

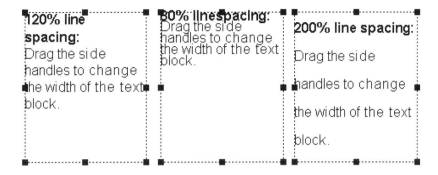

4. Click **Apply** to see the format changes without leaving the dialog box.

5. Click **OK** to dismiss the dialog box.

Changing the Vertical Alignment

Use the following procedure to change the vertical alignment of a text block (text within its alignment box):

1. Select the text.

2. Select **Format | Text**.

3. Click the **Text Block** tab, which controls the formatting of a text block.

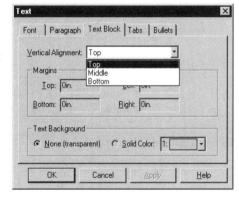

   ➤ The **Top** alignment pushes the text block to the top of its alignment box.

   ➤ The **Middle** alignment pushes the text block to the middle of its alignment box.

   ➤ The **Bottom** alignment pushes the text block to the bottom of its alignment box.

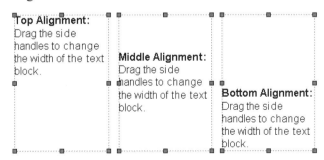

155

4. Click **Apply** to see the format changes without leaving the dialog box.

5. Click **OK** to dismiss the dialog box.

Changing the Margins

Use the following procedure to change the *margin*, which is the distance between a text block and its alignment box:

1. Select the text.

2. Select **Format | Text**.

3. Click the **Text Block** tab, which controls the formatting of a text block.

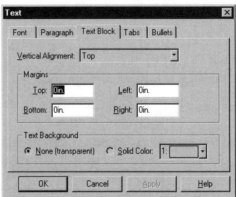

> The **Top** and **Bottom** margins specify the distance between the top (or bottom) of the text block to the alignment box. Type a number to increase the margin distance.

> The **Left** and **Right** margins specify the distance between the left (or right) of the text block to the alignment box. Type a number to increase the margin distance.

4. Click **Apply** to see the format changes without leaving the dialog box.

5. Click **OK** to dismiss the dialog box.

Changing the Background Color

Use the following procedure to change the color of the alignment box, which appears behind the text:

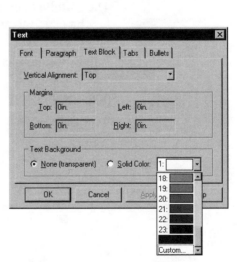

1. Select the text.

2. Select **Format | Text**.

3. Click the **Text Block** tab, which controls the formatting of a text block.

4. Click **None (transparent)** to eliminate color from the alignment box; the default.

5. Alternatively, click **Solid Color**, then select a color. The illustration shows the difference between the two options:

6. Click **Apply** to see the format changes without leaving the dialog box.

7. Click **OK** to dismiss the dialog box.

No Background Color:	Gray Background Color:
Drag the side handles to change the width of the text block.	Drag the side handles to change the width of the text block.

Setting the Tabs

Tabs make it easier to line up columns of text. Press the **Tab** key to move the cursor to the next tab setting. Use the following procedure to set the tab spacing:

1. Select the text.

2. Select **Format | Text**.

3. Click the **Tabs** tab, which controls the tab spacing. Notice that no tabs are initially set.

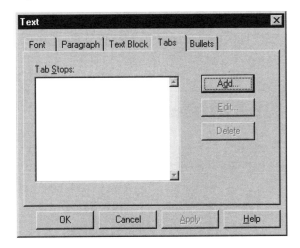

4. Click **Add** to set a tab. Notice the Tab Properties dialog box.

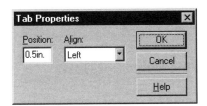

5. Type a number in the **Position** text box.

6. Select an alignment from the **Align** box: left (default), center, right, decimal. The *decimal* tab right-aligns numbers according to the position of their decimal number; it left-aligns text.

7. Click **OK**. Notice the new tab setting in the dialog box.

8. Click **Edit** to change the value of the tab.

9. Click **Delete** to erase the tab.

10. Click **Apply** to see the tab changes without leaving the dialog box.

11. Click **OK** to dismiss the dialog box.

Adding Bullets to Text

A *bullet* is a small symbol that starts a paragraph of text. Using bullets can aid the readability of a list of instructions. A bullet can be any symbol but most commonly is a dot or square.

1. Select the text.

2. Select **Format | Text**.

3. Click the **Bullets** tab. Notice that the dialog box displays seven bullet styles.

No Bullets:
Drag the side handles to change the width of the text block.

- **Square Bullets:**
- Drag the side handles to change the width of the text block.

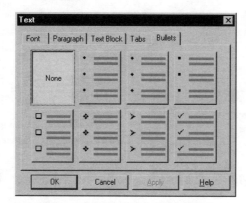

4. Click on the button with the bullet style of your liking.

5. Click **Apply** to see the tab changes without leaving the dialog box.

6. Click **OK** to dismiss the dialog box.

Hands-On Activity

In this activity, you use the text formatting functions. Begin by starting Visio. Then open the **Directional Map** template file found in the **Maps** folder.

1. Drag the **Text block 18pt** shape into the drawing. If necessary, adjust the zoom level to clearly see the text.

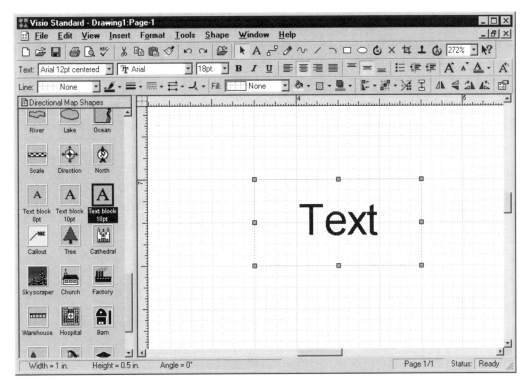

2. Select "Text" by double-clicking it. Notice that the text turns white, with a black highlight.

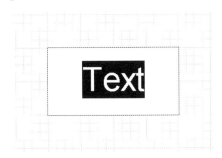

3. Click the **Increase Font Size** button 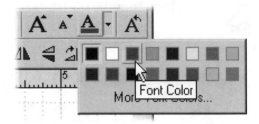 three times. Notice how the word grows larger with each click.

4. Click the **Font Color** button.

5. Select the red color square. Notice that the word turns red.

6. Click the **Bullet** button 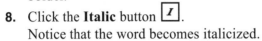. Notice the bullet that prefixes the word.

7. Click the **Bold** button 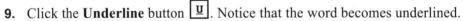. Notice the word becomes bolder.

8. Click the **Italic** button 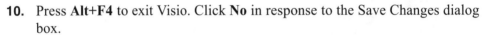. Notice that the word becomes italicized.

9. Click the **Underline** button 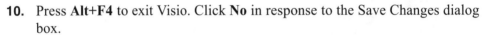. Notice that the word becomes underlined.

10. Press **Alt+F4** to exit Visio. Click **No** in response to the Save Changes dialog box.

This completes the hands-on activity for formatting text in the drawing.

Module 15

Creating Styles
Format | Define Styles, Format Painter

Uses

The **Style** selection of the **Format** menu lets you apply uniform formatting based on the attributes defined by the **Define Styles** selection of the **Format** menu. A *style* is a collection of properties saved by name. In Visio, properties are based on text, line, and fill formats. The advantages to using styles include:

➤ **Consistent look**. When a company defines a corporate look, using styles ensures all illustrators employ the same text, line, and fill formats.

➤ **Faster formatting**. Since a style applies all properties at once, it is faster to apply a single style than to apply each property individually. As Modules 13 and 14 illustrate, shapes and text can have as many as 40 different properties.

➤ **Flexibility**. While applying a style can make everything look similar, Visio gives you the option to override the style with local formatting. *Local* is formatting applied directly.

Visio includes a number of predefined styles; see the composite illustration below. In addition, you can create your own styles. Style definitions are saved with the drawing. You apply a style by selecting it from the **Text**, **Line**, and **Fill** style lists or with the **Style** dialog box.

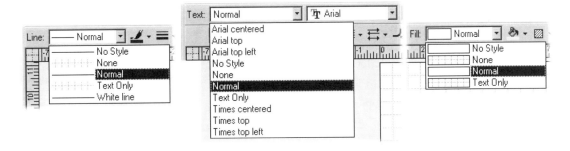

161

The **Format Painter** button of the toolbar lets you copy the formatting of one shape to another.

Procedures

Before presenting the general procedures for defining and using styles, it is helpful to know about the shortcut keys and icons. These are:

Function	Menu	Toolbar Icon
Format Painter		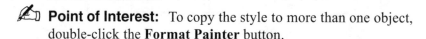
Define Style	Format | Define Styles	
Apply Style	Format | Style	
Text style list box		
Line style list box		
Fill style list box		

Format Painter

Use the following procedure to copy properties from one shape to another:

1. Select the shape with the style you want copied.

2. Click the **Format Painter** button on the toolbar. Notice that the cursor turns into a paintbrush.

> **Point of Interest:** To copy the style to more than one object, double-click the **Format Painter** button.

3. Select another shape to apply the style.

Defining a Style

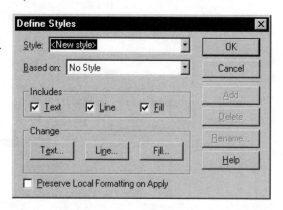

Use the following procedure to create a style:

1. (Optional) To create a named style based on an existing shape, first select the shape.

2. Select **Format | Define Style** from the menu bar. Notice that Visio displays the Define Styles dialog box.

3. Every style must have a name, which is how Visio identifies styles. Type a descriptive name in the **Style** text box. After creating the style, it will appear in the style list boxes on the toolbars.

4. If you want to modify an existing style, first select the style name from the **Based on** list box.

5. Decide whether the style will affect text, lines, and/or fills by selecting the **Text**, **Line**, and **Fill** check boxes.

6. To set the text properties of this style, click the **Text** button. Notice that Visio displays the Text dialog box. Refer to Module 14 for the details of changing text properties.

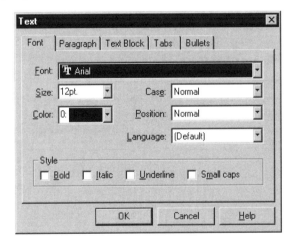

7. To set the line properties of this style, click the **Line** button. Notice that Visio displays the Line dialog box. Refer to Module 13 for the details of changing line properties.

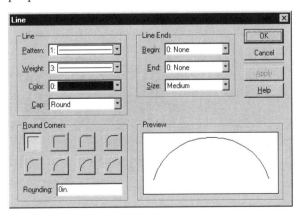

8. To set the fill properties of this style, click the **Fill** button. Notice that Visio displays the Fill dialog box. Refer to Module 13 for the details of changing fill properties.

9. Decide whether you want this style to preserve or override local formatting. To override local formatting, keep **Preserve Local Formatting on Apply** turned off.

✍ **Point of Interest:** *Local* formatting is a change that you make manually. For example, you place some text at the default height of 8 points. Then you change the height to 12 points. The change is a *local* format. Whether you preserve or override local formatting depends on the situation: Sometimes you want to keep local formatting; other times you want to override all those changes.

10. Click **OK** to exit the dialog box.

Applying a Style

Use the following procedure to apply a style:

1. Select one or more shapes.

2. Select **Format | Style** from the menu bar. Notice the Style dialog box.

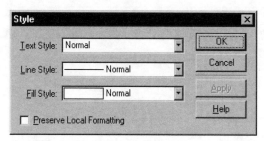

3. Select one or more styles from the three available: **Text Style**, **Line Style**, and/or **Fill Style**. For example, you could select a line and fill style together.

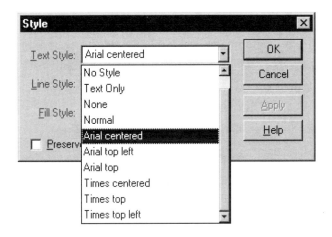

 Point of Interest: The color of the titles Text Style, Line Style, and Fill Style changes, as follows:

Color	Meaning
Black	Style is unchanged from its default setting.
Blue	Style changed prior to opening the dialog box.
Green	Style changed during this dialog box.

When you first open the dialog box, the style titles are black (at their default value) or blue (have been changed at some point). When you change a style, the title turns green.

4. To override local formatting, keep **Preserve Local Formatting on Apply** turned off.

5. Click **Apply** to view the style changes without exiting the dialog box.

6. Click **OK** to exit the dialog box. Notice that the style list boxes on the toolbar update to reflect the selections you made in the Style dialog box.

Text Style List

Use the following procedure to apply a predefined style to text:

1. Select a text block or paragraph in the drawing.

2. Click the **Text** style list box.

3. Select one of the defined style names.

4. Notice that the text changes to match the style.

Line Style List

Use the following procedure to apply a predefined style to a shape:

1. Select a shape in the drawing.
2. Click the **Line** style list box.
3. Select one of the defined style names.
4. Notice that the shape changes to match the style.

Fill Style List

Use the following procedure to apply a predefined style to an area:

1. Select a shape in the drawing.
2. Click the **Fill** style list box.
3. Select one of the defined style names.
4. Notice that the fill changes to match the style.

Hands-On Activity

In this activity, you use the text formatting functions. Begin by starting Visio.

1. Open the template **IDEF0 Diagram Shapes.Vst** found in folder **Flowcharts**.
2. Drag shape **Activity Box** into the drawing. Notice the Custom Properties dialog box.

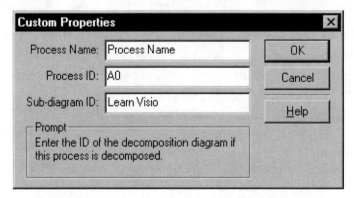

3. Fill in any text for **Sub-diagram ID**, such as "Learn Visio."
4. Click **OK**.
5. If necessary, zoom to make the shape visible.

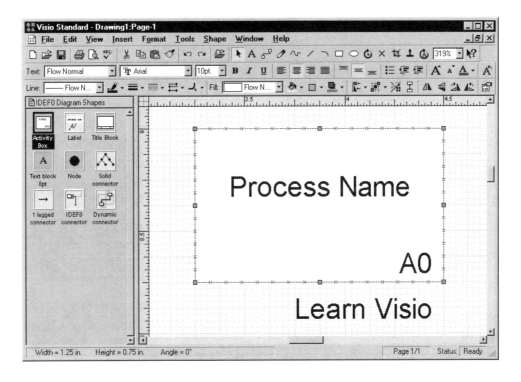

6. Select the shape. Notice the three style list boxes on the toolbar. Each reads, "Flow Normal." This is the name of a predefined style that encompasses text, fill, and line.

7. Click the **Text** style list box. Notice the long list of style names. Most of them are in every Visio drawing. But some were added by the **IDEF0 Diagram Shapes.Vst** template file, such as "Connector" and "Flow 8pt Centered."

8. Select style **Connector** from the **Text** style list box. Notice the warning dialog box Visio displays.

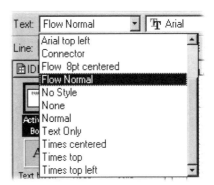

167

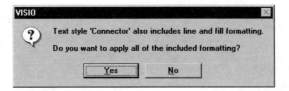

Visio is warning you that the text style also includes style elements for lines and fills. If you click Yes, the existing line and fill styles present in the shape will be overridden.

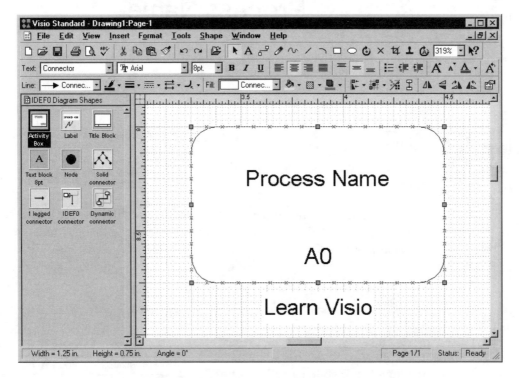

9. Click **Yes**. Notice that the rectangular shape gets rounded corners.

10. Press **Alt+F4** to exit Visio. Click **No** in response to the Save Changes dialog box.

This completes the hands-on activity for using styles in the drawing.

Module 16

Aligning Shapes
Tools | Align Shapes, Distribute Shapes, Array Shapes

Uses

The **Align Shapes**, **Distribute Shapes**, and **Array Shapes** selections of the **Tools** menu automatically rearrange shapes in the drawing.

➤ The **Align Shapes** tool lines up shapes along a horizontal or vertical axis; the tool works with two or more shapes, which are aligned to the first shape selected.

➤ The **Distribute Shapes** tool repositions shapes to create an equal distance between them; the tool works with three or more shapes.

➤ The **Array Shapes** tool creates a rectangular or radial array of shapes. This tool is available only in Visio Technical.

Procedures

Before presenting the general procedures for rearranging shapes, it is helpful to know about the shortcut keys and icons. These are:

Function	Keys	Menu	Toolbar Icon
Align	F8	Tools	Align Shapes	
Distribute		Tools	Distribute Shapes	
Array		Tools	Array Shapes	

Align Shapes

Aligning means that shapes are moved so that they line up vertically or horizontally. Use the following procedure to align several shapes along a baseline:

1. Select at least two shapes. This tool does not work when one or no shapes are selected.

2. This is very important: Notice the color of the selection handles:

Color	Meaning
Green	First selected shape.
Cyan	Other selected shapes.

The first shape you select has green handles; the ensuing shapes have cyan (light blue) handles. The **Align** tool aligns the ensuing shapes to the first shape.

3. Select **Tools | Align Shapes**. Notice the Align Shapes dialog box.

4. Select the type of alignment (you can select either or both types of alignment):

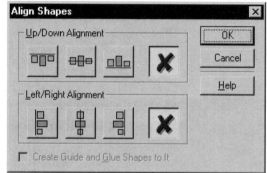

> **Up/Down Alignment:** Align the top, center, or bottom of the shapes to the first shape selected.

> **Left/Right Alignment:** Align the left, center, or right of the shapes to the first shape selected.

5. Click **Create Guide and Glue Shapes to It** when you want Visio to create a guide line along the alignment axis; the selected shapes will be glued to the guide line.

6. Click **OK**. Notice that Visio moves the shapes into alignment. It is possible for some shapes to "disappear." What has happened is that a larger shape has covered up a smaller shape.

The following illustration shows unaligned shapes (left) and top-aligned shapes glued to a horizontal guide (left). The "No" symbol was the first shape selected, so the other two shapes aligned to the top of its alignment box.

Distribute Shapes

Distribute means that shapes are moved so that they are evenly spaced from each other. For this reason, you must select three or more shapes before Visio lets you use this tool. The two outer shapes remain stationary, while the center shape(s) move to create the even spacing. Use the following procedure to distribute shapes:

1. Select at least three shapes.

2. Select **Tools | Distribute Shapes**. Notice the Distribute Shapes dialog box.

3. Select one type of distribution.

4. Click **Create Guides and Glue Shapes to Them** when you want Visio to create a guide line along the alignment axis; the selected shapes will be glued to the guide line.

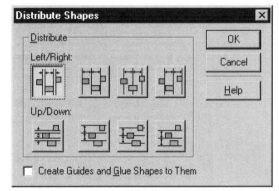

5. Click **OK**.

6. The middle shape(s) move so that the spacing is equal between the shapes. Notice that the outer shapes (on the left and right side) remain in place. In the illustration, three shapes were distributed horizontally, with three vertical guide lines.

Array Shapes (Visio Technical Only)

Array means to create a rectangular pattern of shapes from a selection of one or more shapes. This tool is available only in Visio Technical. Use the following procedure to array a shape into a rectangular pattern:

1. Select one or more shapes.

2. Select **Tools | Array Shapes**. Notice the Array Shapes dialog box.

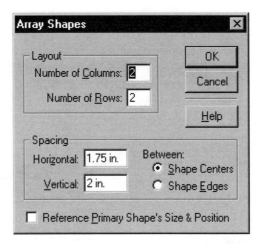

➤ **Layout::** Type the number of rows and columns for the array (default = 2).

➤ **Spacing:** Type the **Horizontal** and **Vertical** distances for the space between shapes. A positive distance arrays the shapes to the right and bottom; a negative distance arrays the shapes to the left and top. Default = the shape's size.

➤ **Between:** Select whether the spacing is between the **Centers** or the **Edges** of the shape. When you select **Centers** and the spacing is unchanged, the shapes are placed immediately next to each other. When you select **Edges** (and the spacing is unchanged), the shapes have a distance equal to their size between each other. The illustration shows centers (left) and edges (right):

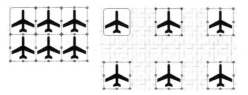

➤ **Reference Primary Shape's Size & Position:** Visio draws guide lines through the arrayed shape's center or edge.

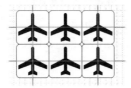

3. Click **OK**. Notice that Visio creates a rectangular array of shapes.

Hands-On Activity

In this activity, you use the align, distribute, and array tools. Begin by starting Visio.

1. Open the stencil **Symbols.Vss** found in folder **Visio Extras**.
2. Drag the following shapes into the drawing: Yen, Rail Transportation, and Lightning.
3. Place the shapes anywhere in the drawing, making sure they are not aligned!
4. Adjust the level of zoom to see the shapes clearly.

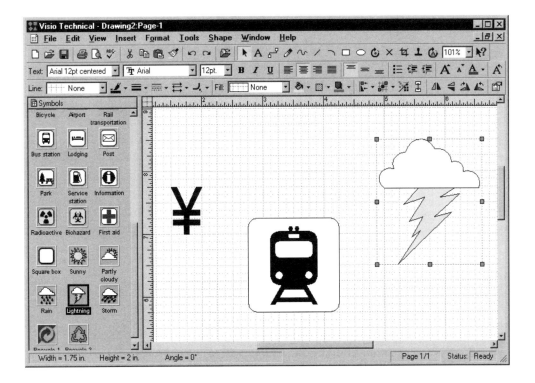

5. Select the lighting shape.
6. Hold the **Shift** key, then select the other two shapes.

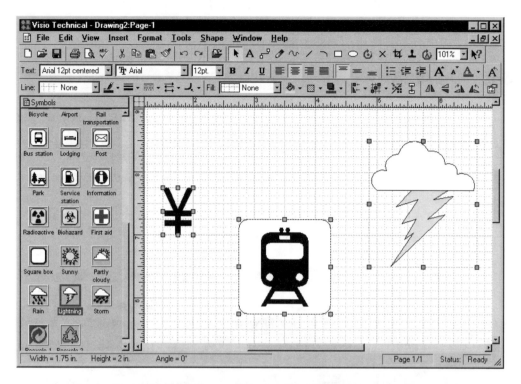

7. Select **Tools | Align Shapes**. Notice the **Align Shapes** dialog box.

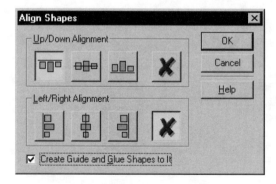

8. Select the first button (align with top edge) in **Up/Down Alignment**.
9. Click **Create Guide and Glue Shapes to It**.
10. Click **OK**.
11. Notice how Visio moves the two shapes you selected with the **Shift** key. Their top edges line up with the star's top edge. In addition, Visio glues the three shapes to the guide line, which Visio placed along the top edge.

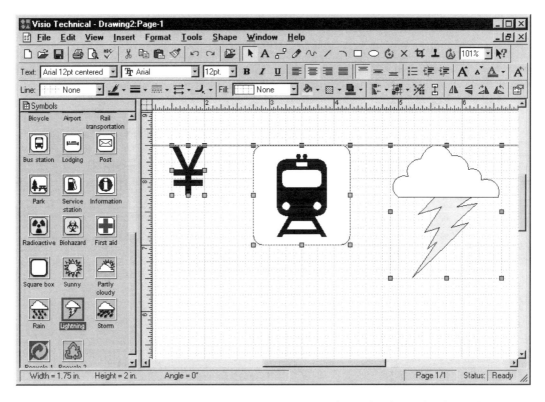

12. Select the guide line and move it around. Notice how the three shapes move along, because they are glued to the guide line.

13. Ensure the three shapes are selected.

14. Select **Tools | Distribute Shapes**. Notice the Distribute Shapes dialog box.

15. Select the first button (align center shapes left and right).

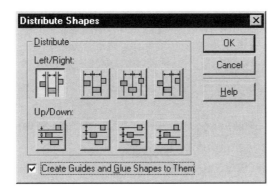

16. Click **Create Guides and Glue Shapes to Them**.

17. Click **OK**.

18. Notice how Visio moves the center shape to be equidistant between the two outer shapes. In addition, a vertical guide line is placed through the center of each symbol.

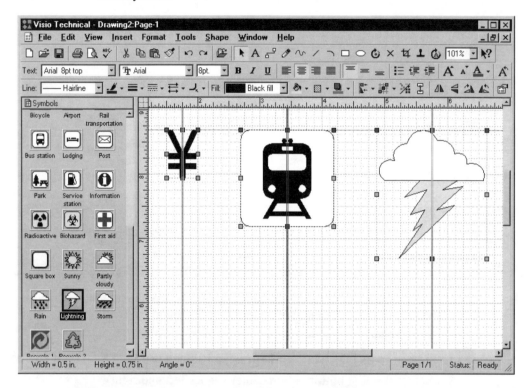

The following steps work only with Visio Technical:

19. Press **Ctrl+A** to select everything in the drawing.

20. Press **Delete** to erase everything in the drawing.

21. Drag the **Information** shape into the drawing.

22. Make sure the shape is selected, then select **Tools | Array Shapes** from the menu bar.

23. In the **Layout** array, type **3** for **Number of Columns**.

24. Type **2** for **Number of Rows**.

25. Leave the **Spacing** at 1.5.

26. Click the check box next to **Reference Primary Shape's Size & Position**.

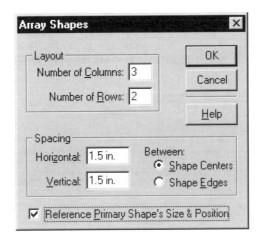

27. Click **OK**.

28. Notice that Visio creates a 3-by-2 array of six shapes with horizontal and vertical guide lines.

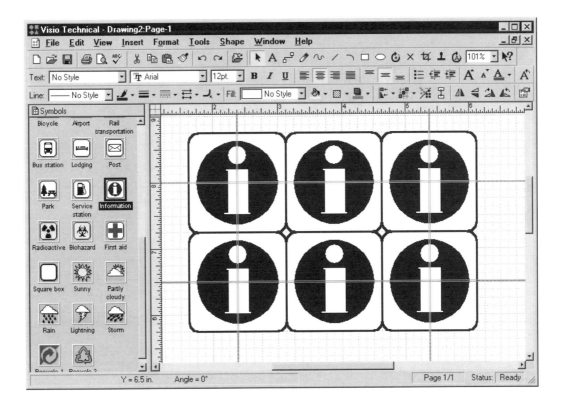

29. Press **Alt+F4** to exit Visio. Click **No** in response to the Save Changes dialog box.

This completes the hands-on activity for aligning, distributing, and arraying shapes in the drawing.

Module 17

Creating Groups

Shape | Grouping

Uses

The **Group** function, found on the **Shape** menu, is used to group shapes together into a single unit. A group contains shapes and other groups. Once in a group, editing commands apply equally to all members of the group. For example, when you select the color red for lines, all shapes in the group get red lines.

You can **Add** and **Remove** shapes from a group. You can **Convert** some non-Visio objects to a group, such as an image placed from another program.

Many shapes that you drag into the drawing from a stencil are groups. To edit individual lines, you **Ungroup** the grouped shape. However, ungrouping a stencil shape causes the shape to lose its link to the stencil. Proceed with care when ungrouping.

Procedures

Before presenting the general procedures for grouping, it is helpful to know about the shortcut keys and icons. These are:

Function	Keys	Menu	Toolbar Icon
Group	Ctrl+G	Shape | Grouping | Group	🔲
Add		Shape | Grouping | Add to Group	
Remove		Shape | Grouping | Remove from Group	
Convert		Shape | Grouping | Convert to Group	
Ungroup	Ctrl+U	Shape | Grouping | Ungroup	🔲

Create a Group From Shapes

Use the following procedure to make a group:

1. Select one or more shapes.
2. Select **Shape | Grouping | Group**.
3. Notice that Visio indicates the group status by placing a single set of handles around all shapes.

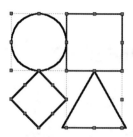

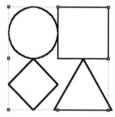

Add to a Group

Use the following procedure to add shapes to a group:

1. Select a group and one or more shapes.

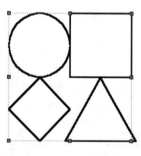

2. Select **Shape | Grouping | Add to Group**.
3. Notice that Visio indicates the group status by placing a single set of handles around all shapes.

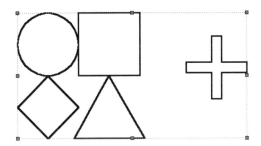

Remove From a Group

Use the following procedure to remove a shape from a group:

1. Select one or more shapes within a group. Notice the shapes have gray-colored handles.
2. Select **Shape | Grouping | Remove from Group**.
3. Notice that Visio indicates the group status by placing a single set of handles around all shapes.

 Point of Interest: When working with groups, the color of the handles is important:

Color	Meaning
Green	First selected shape.
Cyan	Subsequent selected shapes.
Gray	Shape is part of a group.

When you first select a shape, it has green handles. Hold down the **Shift** key to select additional shapes whose handles are cyan (light blue). However, when the shape is part of a group, the handles are gray—no matter the order in which you select the shape.

Disband a Group

Use the following procedure to ungroup shapes:

1. Select a group.
2. Select **Shape | Grouping | Ungroup**.
3. Notice that Visio indicates the removal of group status by placing handles around all shapes within the group.

4. If the group contains nested groups, you may need to apply the **Ungroup** command a second time.

 Point of Interest: The **Remove** and **Ungroup** tools have a different effect when working with existing groups. When you apply the **Remove** command to a group, you remove a shape; the remaining shapes remain a group.

When you apply the **Ungroup** command to a group, all shapes are removed from the group. The group no longer exists.

Hands-On Activity

In this activity, you use the grouping functions. Begin by starting Visio.

1. Open a new drawing, along with the **Basic Shapes.Vss** stencil found in the **Block Diagram** folder.

2. Drag the **Triangle**, **Square**, and **Pentagon** shapes into the drawing.

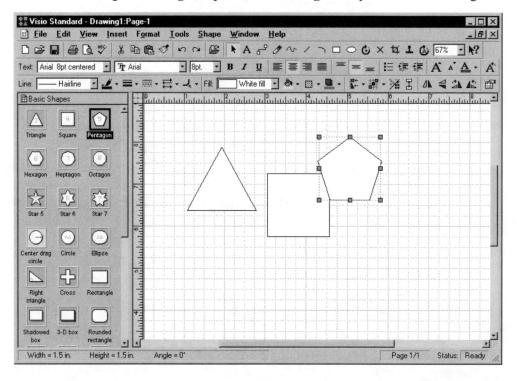

3. Select a shape.

4. Select the other two shapes by holding down the **Shift** key.

5. Select **Shape | Grouping | Group** from the menu bar. Notice that one set of handles surrounds all three shapes.

6. Give the shapes a heavier line weight. Notice that the change applies to all members of the group.

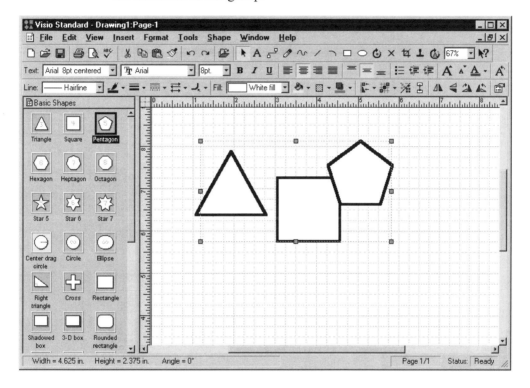

7. Select the pentagon shape. Notice the gray handles that surround it.

8. Attempt to increase the size of the pentagon by dragging one of the handles. Notice that you cannot.

9. Give the pentagon a red fill color. Notice that the change happens.

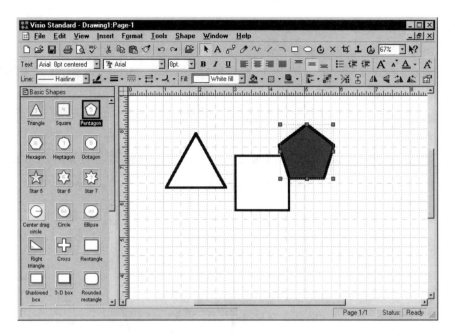

10. Select the group.

11. Drag a corner to resize the group. Notice that you can change the size of the group but not of members in the group.

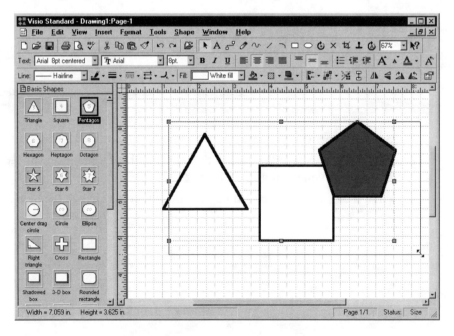

12. Select **Shape | Grouping | Ungroup**. Notice that each shape has its own set of handles again.

13. Press **Alt+F4** to exit Visio. Click **No** in response to the Save Changes dialog box.

This completes the hands-on activity for working with groups in the drawing.

Module 18

Boolean Operations

Shape | Operations
Tools | Macro | Maps | Build Region

Uses

The **Operations** selection of the **Shape** menu performs Boolean operations on shapes. You use Boolean operations to create new shapes. Very often, the order in which you select shapes is important to the outcome, as noted below. For operations to work, the shapes must be overlapping. The stacking order of overlapping shapes is *not* important.

➤ The **Union** operation joins all selected shapes into a single shape. The new shape takes on the attributes of the shape selected first.

➤ The **Combine** operation is like the **Union** operation but removes the portions in common. The new shape takes on the attributes of the shape selected first.

➤ The **Fragment** operation creates three new shapes from two overlapping shapes; the overlapping portion becomes an independent shape. All three shapes take on the attributes of the first selected shape.

➤ The **Intersect** operation removes everything *except* the overlapping areas of the two shapes. **Intersect** is the opposite of the **Combine** operation.

➤ The **Subtract** operation removes the overlapping portion of the second shape from the first shape. Another way of looking at it is that the overlapping portion is removed from the first shape.

➤ The **Build Region** macro joins shapes that have been enabled for arranging. Specifically, the macro joins map shapes, such as countries and states.

Procedures

Before presenting the general procedures for Boolean operations, it is helpful to know about the shortcut keys. These are:

Function	Menu
Union	Shape | Operations | Union
Combine	Shape | Operations | Combine
Fragment	Shape | Operations | Fragment
Intersect	Shape | Operations | Intersect
Subtract	Shape | Operations | Subtract
Build Region	Tools | Macro | Maps | Build Region

Union Operation

Use the following procedure to join two or more shapes together:

1. Select one shape. This is the shape whose properties the new shape takes on.

2. Hold down the **Shift** key and select one or more additional shapes.

3. Select **Shape | Operations | Union**.

4. Notice that Visio creates one new shape with the outline of all selected shapes and the attributes of the first shape.

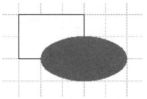

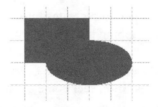

Combine Operation

Use the following procedure to join two or more shapes together, then subtract the areas in common:

1. Select one shape. This is the shape whose properties the new shape takes on.

2. Hold down the **Shift** key and select one or more additional shapes.

3. Select **Shape | Operations | Combine**.

4. Notice that Visio creates one new shape with the
    outline of all selected shapes and the attributes of
    the first shape. Overlapping areas are removed.

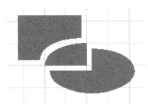

Fragment Operation

Use the following procedure to create three or more
shapes from two or more shapes, where overlapping
portions become independent shapes:

1. Select one shape. This is the shape whose properties the new shape takes on.

2. Hold down the **Shift** key and select one or more additional shapes.

3. Select **Shape | Operations | Fragment**.

4. Notice that Visio creates a new fragmented shape
    from the overlapping portions of the original
    shapes. All shapes take on the attributes of the first
    shape. The illustration shows the three shapes
    moved apart.

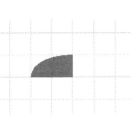

Intersect Operation

Use the following procedure to create one shape from
the overlapping portion of two or more shapes:

1. Select one shape. This is the shape whose properties the new shape takes on.

2. Hold down the **Shift** key and select one or more additional shapes.

3. Select **Shape | Operations | Intersect**.

4. Notice that Visio creates one new shape from the
    overlapping portions of the original shapes; the
    new shape takes on the attributes of the first shape.

Subtract Operation

Use the following procedure to subtract one shape from
another shape:

1. Select one shape. This is the shape whose proper-
    ties the new shape takes on.

2. Hold down the **Shift** key and select one or more additional shapes.

3. Select **Shape | Operations | Subtract**.

4. Notice that Visio creates one new shape by removing the second shape and the overlapping portions of the second shape from the first shape.

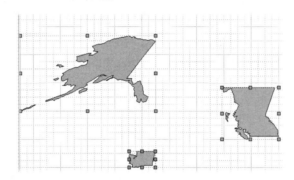

Building a Region

Use the following procedure to build a region from map shapes:

1. Select two or more map shapes.

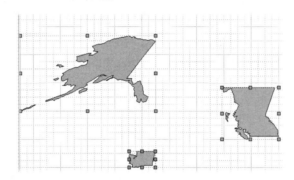

2. Select **Tools | Macro | Maps | Build Region** from the menu bar.

3. Notice that Visio moves the map shapes together so that they join up logically along borders. In the illustration, the American states of Alaska and Washington, and the Canadian province of British Columbia, have been brought together by the **Build Region** command.

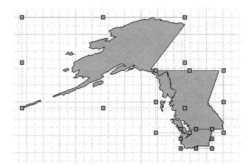

Hands-On Activity

In this activity, you use the Boolean operations. Begin by starting Visio.

1. Open a new document.
2. Open stencil file **Basic Shapes.Vss** from folder **Block Diagram**.
3. Drag the **Triangle** shape into the drawing.
4. Drag the **Circle** flag shape into the drawing, half overlapping the triangle.
5. Apply the diagonal line fill pattern to the circle.

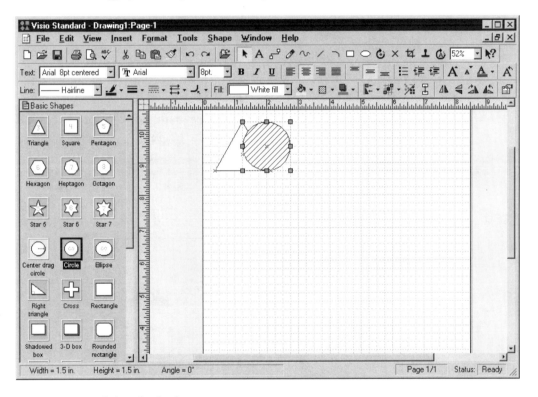

6. Select both shapes.
7. With both shapes still selected, hold down the **Ctrl** key to make five copies of the originals. Remember to use the **Shift** key to force the movement to horizontal and vertical. Use the **F4** key to repeat an action.
8. Select the first circle, then the triangle. Remember to hold down the **Shift** key when selecting the triangle.

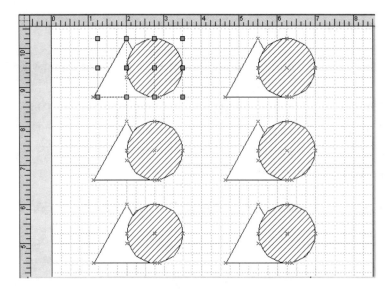

9. Select **Tools | Operations | Union**. Notice that the two shapes become a single shape and take on the fill pattern of the circle.

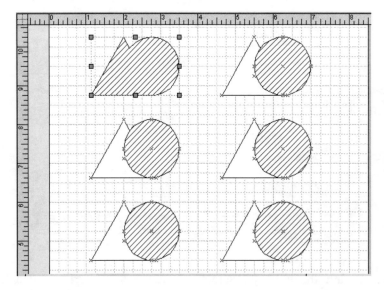

10. Select the second circle, then the triangle.

11. Select **Tools | Operations | Combine**. Notice that the two shapes become a single shape and take on the fill pattern of the circle—except for the area in common.

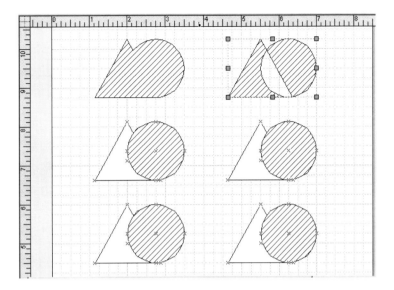

12. Select the third circle, then the triangle.

13. Select **Tools | Operations | Fragment**. Notice that the two shapes become three shapes and take on the fill pattern of the circle. The illustration shows the three shapes moved apart.

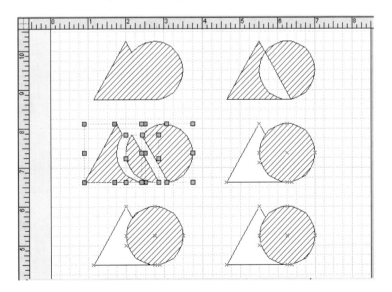

14. Select the fourth circle, then the triangle.

15. Select **Tools | Operations | Intersect**. Notice that the two shapes become a single shape—consisting of the area in common—and take on the fill pattern of the circle.

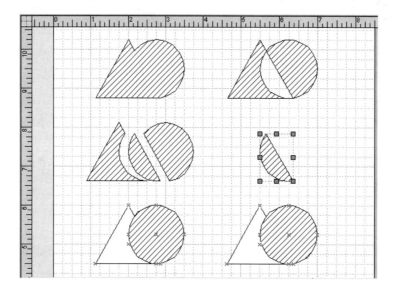

16. Select the fifth circle, then the triangle.

17. Select **Tools | Operations | Subtract**. Notice that the triangle is removed from the circle.

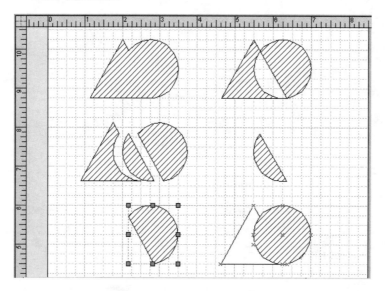

18. To show the importance of selection order, this time select the sixth triangle, then the circle.

19. Select **Tools | Operations | Subtract**. Notice that the circle is removed from the triangle. Since the triangle was selected first, the resulting Boolean shape takes on the attributes of the triangle.

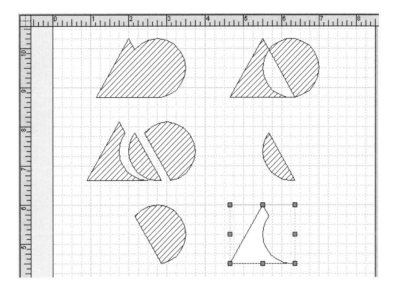

20. Press **Alt+F4** to exit Visio. Click **No** in response to the Save Changes dialog box.

This completes the hands-on activity for Boolean operations. Being familiar with the five operations makes it easier to create certain kinds of new shapes.

Module 19

Previewing Before Printing

File | Print Preview

Uses

The **Print Preview** selection of the **File** menu lets you see how the drawing will appear on the paper, before committing to printing. By previewing the print, you ensure the drawing will appear at the correct size and orientation on the paper.

Procedures

Before presenting the general procedure for print preview, it is helpful to know the shortcut key. It is:

Function	Keys	Menu	Toolbar Icon
Preview		File | Print Preview	

Print Preview

Use the following procedure to preview the drawing before printing:

1. Select **File | Print Preview**.
2. Notice that Visio displays the drawing on a white background, representing the sheet of paper. The light gray edging represents the printer margins; if any portion of the drawing extends into the margin, that part will not be printed.

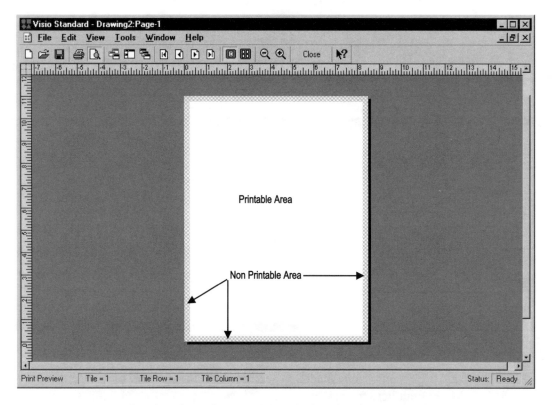

3. Notice that the **Print Preview** window has its own toolbar:

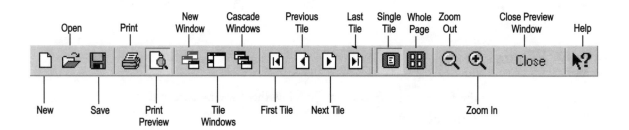

> ➤ **New**, **Open**, and **Save**: You can open another Visio drawing from Print Preview mode. However, the drawing is opened in Visio, not in Print Preview.

> ➤ **Print** and **Print Preview**: The Print button displays the Print dialog box. Clicking the Print Preview button is like clicking the Close button: both take you back to the regular Visio window.

➤ **New Window**, **Tile Windows**, and **Cascade Windows**: You can have a print preview of more than one drawing at a time. These commands let you tile the windows to see two or more drawings side by side.

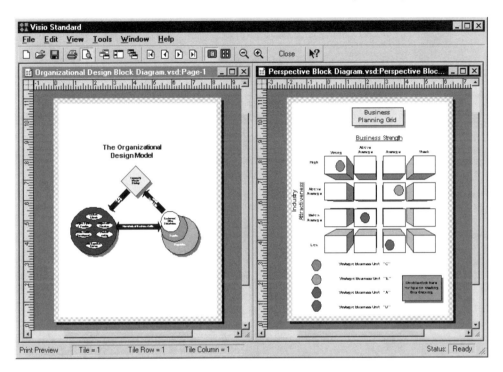

➤ **Single Tile** and **Whole Page**: When the drawing is larger than one sheet of paper, Visio *tiles* the drawing. Tiling means that the drawing will be printed on enough sheets of paper to print the full drawing, which you can tape together. Single Tile shows one sheet of a multitile printout; Whole Page shows all the sheets.

➤ **First Tile**, **Previous Tile**, **Next Tile**, and **Last Tile**: These commands work only in **Single Tile** mode. They take you from tile to tile. The status line reports which tile you are looking at. For example:

Tile = 2   Tile Row = 1   Tile Column = 2

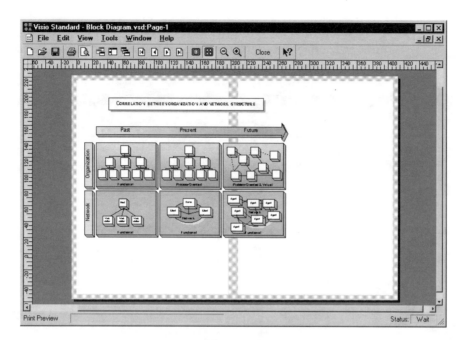

➤ **Zoom Out** and **Zoom In**: The default tool is **Zoom**. Click the drawing to examine it more closely. Visio performs a 2:1 zoom. Click the drawing to return back to the full-page view.

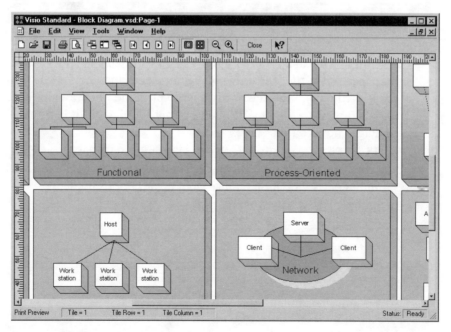

> ➤ **Close** and **Help**: **Close** takes you back to the Visio drawing screen, which it calls "normal view." If you have more than one print preview window, all are closed. The **Help** button brings up context sensitive help.

5. Select **File | Print** to print the drawing.

6. Or, click the **Close** button on the toolbar to return to the drawing without printing.

Hands-On Activity

In this activity, you use the print preview function. Begin by starting Visio.

1. Select **File | New | Browse Sample Drawings**.

2. From the **Browse Sample Drawings** dialog box, go to the **Block Diagram** folder and double-click **Perspective Block Diagram.Vsd**. Notice that Visio loads it.

3. Select **File | Print Preview**. Notice that Visio displays the drawing in a window with a gray background.

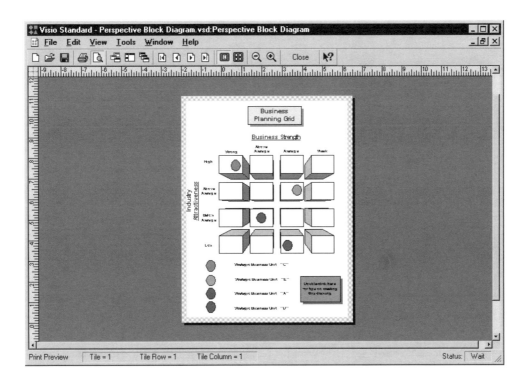

4. Move the cursor over the drawing. Notice the cursor is a magnifying glass with a + sign. This indicates Visio is in zoom mode.

5. Click the drawing. Notice that Visio enlarges the view.

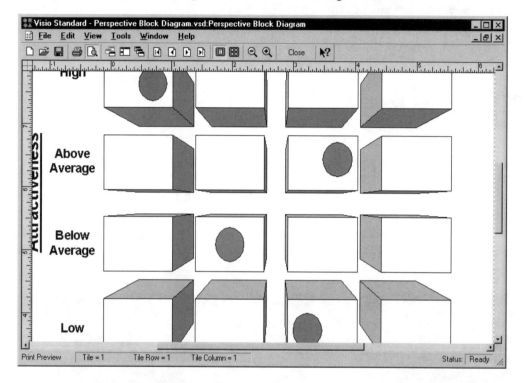

6. Notice that the cursor's + sign has changed to a – sign. This indicates Visio is ready to zoom out to the full-page view.

7. Click the drawing.

8. Notice Visio reduces the view.

9. Click **Close** on the toolbar to return to the drawing.

10. Press **Alt+F4** to exit Visio.

This completes the hands-on activity for print preview.

Module 20

Printing Drawings

File | Print

Uses

The Print selection on the File menu prints the drawing. Under Windows, you are not limited to printing to one particular printer. For example, if your computer is connected to a fax-modem, you can send the drawing as a fax. If your computer is hooked up to an e-mail system, you can attach the drawing to an e-mail message with the File | Send command.

Procedures

Before presenting the general procedures for printing, it is helpful to know about the shortcut keys. These are:

Function	Keys	Menu	Toolbar Icon
Print	Ctrl+P	File	Print	
Email		File	Send	

✍️ **Point of Interest:** The Print button works different from the File | Print command. Clicking the Print button causes Visio to immediately print the drawing, without displaying a dialog box. Selecting File | Print displays the Print dialog box.

Printing the Drawing

Use the following procedure to print the drawing:

1. Select **File | Print**. Notice the Print dialog box.

2. Select the name of the printer from the **Name** list box. To fax the drawing, select the **Microsoft Fax** or other fax "printer" from this list.

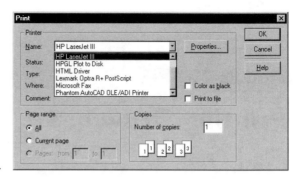

3. Click the **Properties** button to change the characteristics of the printer, such as resolution and paper orientation.

4. Click **OK** to return to the Print dialog box.

5. Click **Color as black** to ensure all shapes are printed on a monochrome printer (one that is incapable of printing color or shades of gray).

6. Click **Print to file** if you want the drawing saved in a file. This option usually only makes sense as a translator, such as converting the file to HPGL (short for Hewlett-Packard graphics language), which can be imported into another drawing or CAD program.

7. If the drawing consists of more than one page, select which pages you want printed in the **Page range** area:
   - ➤ **All:** Prints all pages in the drawing (the default).
   - ➤ **Current Page:** Prints the currently displayed page.
   - ➤ **Pages from:** When your drawing consists of one page, this option is unavailable.

8. Type the number of copies in the **Number of copies** text box.

9. Click **OK** to begin printing.

10. Alternatively, click **Cancel** to dismiss the dialog box without printing the drawing.

Hands-On Activity

In this activity, you use the print function. Ensure Visio is running.

1. Select **File | New | Browse Sample Drawings**.

2. Notice the Browse Sample Drawings dialog box. From the Flowchart folder, double-click the **Hoshin Flowchart.Vsd** drawing file.

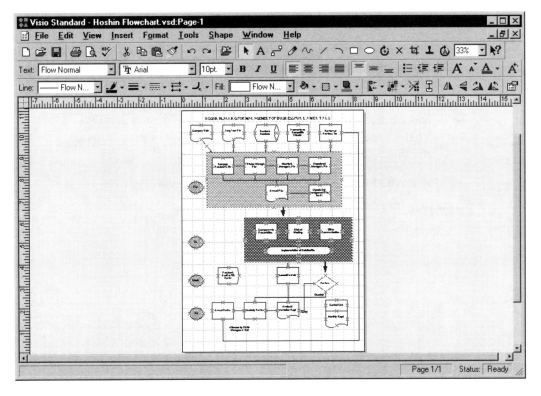

3. Select **File | Print**.

4. Notice the Print dialog box.

5. Click **OK**.

6. Notice the Printing dialog box, which tells you of Visio's progress in sending the drawing to the printer.

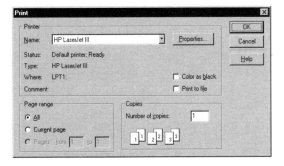

7. Wait for the printer to print the drawing.

 Point of Interest: Most Windows systems are set up to use the Print Manager. This allows a software application, like Visio, to finish printing sooner. However, you will probably have to wait for a minute or two after Visio finishes printing. That's because the printer is usually slower than the computer running Visio.

8. Press **Alt+F4** to exit Visio.

This completes the hands-on activity for printing.

Module 21

Undoing and Redoing

Edit | Undo, Redo

Uses

The **Undo** and **Redo** selections of the **Edit** menu reverse your work. When you make a mistake, such as deleting a shape, **Undo** returns the erased shape. Visio remembers the last ten actions you performed; repeating the **Undo** command reverses each action, one at a time. To help you remember what the next undo action will be, the **Edit** menu follows **Undo** with the action, such as **Undo Edit Segment**.

➤ Some actions cannot be undone, such as saving, printing, and certain shape operations. When the action cannot be undone, the **Edit** menu shows **Can't Undo** instead of **Undo**.

➤ If you change your mind a second time, the **Redo** command undoes the **Undo**. Unlike **Undo**, Visio remembers a single **Redo** action. Naturally, you can undo a redo.

Procedures

Before presenting the general procedures for undoing and redoing, it is helpful to know about the shortcut keys. These are:

Function	Keys	Menu	Toolbar Icon
Undo	Ctrl+Z	Edit	Undo	↰
Redo	Ctrl+Y	Edit	Redo	↱

Undoing an Action

Use the following procedure to undo an action:

1. Select **Edit | Undo** or press **Ctrl+Z**.

   2. If necessary, select **Edit | Undo** again.

Redoing an Undo

   To redo the previous undo, select **Edit | Redo** or press **Ctrl+Y**.

Hands-On Activity

   In this activity, you use the undo and redo functions. Begin by starting Visio.
   Then start a new document with the **Basic** template:

   1. Drag the 5-point **Star 5** shape from the **Basic** stencil to the center of the page.

   2. Drag the **Square** shape from the stencil to a location above the star.

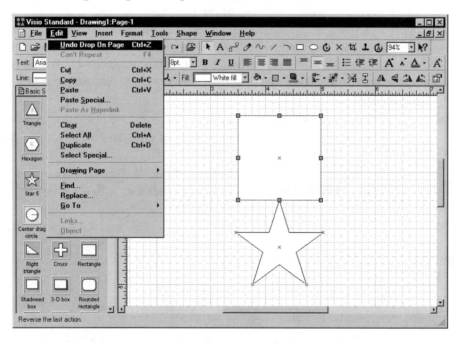

   3. Select **Edit | Undo Drop On Page**. Notice that the square disappears from the page.

   4. Drag the **Circle** shape from the stencil to on top of the **Star**.

   5. Select **Gray fill** from the **Fill** list box to turn the circle gray.

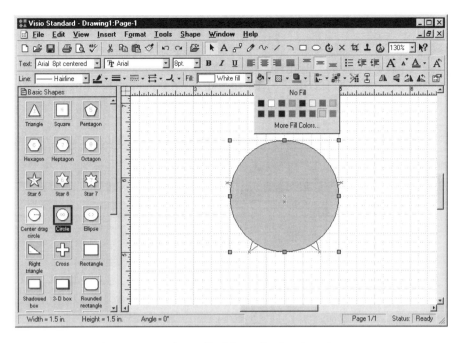

6. Press **Ctrl+A** to select all objects in the drawing.

7. Select **Shape | Operations | Combine.** Notice that the star shape is "punched" out of the gray circle.

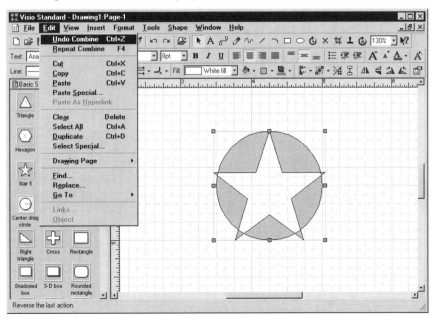

8. Select **Edit | Undo Combine**. Notice that the circle reverts to its original form.

9. Select **Edit | Undo Fill Style**. Notice that the circle loses its gray color.

10. Select **Edit | Redo Fill Style**. Notice that the circle turns gray again.

11. Select **Edit | Redo Combine**. Notice that the star shape is removed from the circle.

12. Press **Alt+F4** to exit Visio. Click **No** in response to the Save Changes dialog box.

This completes the hands-on activity for undo and redo operations.

Module 22

Help

Help | Visio Help Topics

Uses

The **Help** menu displays a condensed version of the documentation. When you select **Help | Visio Help Topics**, you can search for documentation via the **Contents**, the **Index**, and the **Find** function. The **Contents** lists help topics in logical order; the **Index** lists topics in alphabetical order; the **Find** function searches for specific keywords.

As an alternative, you can click the **Help** button on the toolbar, then click anywhere on the Visio window: toolbar buttons, menu selections, and the drawing area.

Procedures

Before presenting the general procedures for help, it is helpful to know about the shortcut key:

Function	Keys	Menu	Toolbar Icon
Help	FI	Help	Visio Help Topics	▶?
What's This?		Shift+FI Help	What's This	

Hands-On Activity

Practice using Help. Begin by starting Visio with no document open. Use the following procedure to search for help:

1. Select **Help | Visio Help Topics**.

2. Notice that Visio displays the Help Topics dialog box.

3. In the **Contents** tab, double-click on the book and page icons.

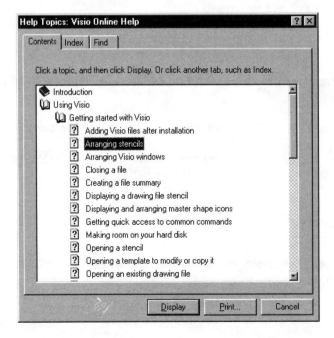

4. Notice that Visio displays the documentation associated with the help topic.

5. Notice the green underlined text. This text is linked to other help topics. When the underlining is a dashed green line, the word has a definition attached. Click on a green underlined text item and read the definition.

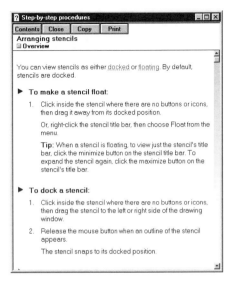

Docked stencil
A stencil that is attached to the side of the drawing window. By default, stencils are docked on the left side of the window. You can make stencils float, or you can dock them on the right side of the drawing window. See also floating stencil.

6. Click anywhere on the screen to dismiss the definition box.

7. Click the **Contents** button to return to the Contents dialog box.

8. Click the **Index** tab to display the index window.

9. Type the name of a Visio topic in the **1. Type the first few letters of the word you're looking for** text box.

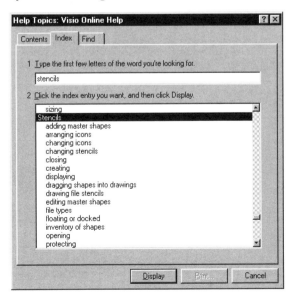

10. Click the phrase that is closest to the topic you are searching for.

11. Click **Display**. Notice that Visio displays the Topics Found dialog box.

12. Click on the topic phrase that best matches the topic you are looking for.

13. Click **Display**. Notice that Visio displays the related help topic in the Step-by-step procedures dialog box.

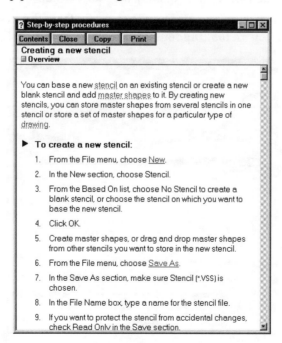

14. Click the **Contents** button to return to the Contents dialog box.

15. Click the **Find** tab to display the Find dialog.

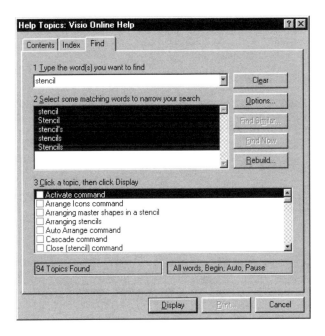

16. Type the word you want to find in the **1** text box.

17. Narrow your search by selecting one or more words in the **2** list box.

18. Select the help topic in the **3** list box.

19. Click **Display**. Notice that Visio displays the related help topic in the Online Help dialog box.

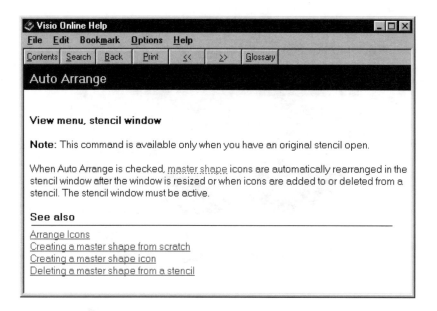

📝 **Point of Interest:** The Step-by-step procedures dialog box is a special type of dialog box that always floats on top of all other windows on your screen. That means it could obscure other help windows. When done with a Step-by-step procedures dialog box, click **Close**.

20. Click **File | Exit** to dismiss the help window.

This completes the procedures for using help.

Module 23:

Drawing Tools

Pencil, Line, Arc, Freeform, Rectangle, Ellipse

Uses

 The drawing tools are found on the toolbar—not on any menu. They draw straight and curved lines, circles and arcs, rectangles, and ellipses. Many of Visio's drawing tools are dual purpose: A different object is created depending on how you move the mouse or hold down a key.

Tool	Draws	Special Action
Pencil	Straight line	Circular arc: move mouse in a curve.
		Line at 45-degree increments: hold down **Shift** key.
Line	Straight line	Line at 45-degree increments: hold down **Shift** key.
Arc	Elliptical arc	Line: move mouse in a horizontal or vertical direction.
Rectangle	Rectangle	Square: hold down **Shift** key.
Ellipse	Ellipse	Circle: hold down **Shift** key.
Freeform	NUBS curves	

The only tool that isn't dual purpose is the **Freeform** tool (also called the **Spline** tool), which draws NUBS curves. *NUBS* is short for nonuniform Bezier spline and should not be confused with the more common NURBS curve (nonuniform *rational* Bezier spline) found in computer-aided design software.

Another way of looking at the drawing tools is to list them by the objects they draw:

To draw a	Use this tool
Line	Line tool.
	Pencil tool: move mouse in a straight line.
Constrained line	Line tool: hold down **Shift** key.

To draw a	Use this tool
	Pencil tool: hold down **Shift** key.
	Arc tool: move mouse in a horizontal or vertical direction.
Circular arc	Arc tool.
Curve	Freeform tool.
Circle	Ellipse tool: hold down **Shift** key.
Elliptical arc	Pencil tool: move mouse in a curve.
Ellipse	Ellipse tool.
Rectangle	Rectangle tool.
Square	Rectangle tool: hold down **Shift** key.
Spline	Freeform tool.

 Point of Interest: So that you know where an open object, such as a line or arc, starts and ends, Visio uses two symbols:

Symbol	Meaning
x	Starting point of a line or arc.
+	Ending point.

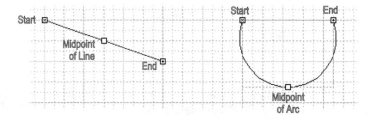

After you draw a line or arc, notice that there is a small x where you began drawing and a small + where you stopped drawing. To connect one line or arc to another, start drawing from the endpoint. Visio automatically connects the two.

Procedures

Before presenting the general procedures for drawing, it is helpful to know the shortcut keys. These are:

Function	Keys	Toolbar Icon
Pencil Tool	Ctrl+4	
Line Tool	Ctrl+6	
Arc Tool	Ctrl+7	
Spline Tool	Ctrl+5	
Rectangle Tool	Ctrl+8	
Ellipse Tool	Ctrl+9	

Drawing with the Pencil Tool

Use the following procedure to draw a line or a circular arc with the **Pencil** tool:

1. Click **Pencil tool**.
2. Drag (hold down the button and move the mouse) to draw with the Pencil tool.
3. Move the mouse in a straight line to draw a straight line at any angle.
4. Hold down the **Shift** key to draw a line in 45-degree increments.
5. Or, slowly move the mouse in the shape of an arc and the Pencil tool draws a circular arc. You cannot draw a circle with this tool.

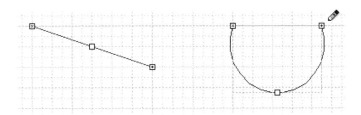

 Point of Interest: To draw a line in *45-degree increments* means that Visio draws the line at one of the 45-degree angles—0, 45, 90, 135, 180, 225, 270, and 315 degrees—nearest to the angle your mouse is moving. The illustration shows lines drawn at 45-degree increments.

Drawing with the Line Tool

Use the following procedure to draw a line with the **Line** tool:

1. Click the **Line tool**.
2. Drag to draw a line at any angle with the Line tool.
3. Hold down the **Shift** key to draw a line in 45-degree increments.

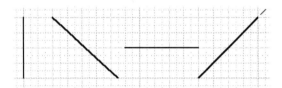

Drawing with the Arc Tool

Use the following procedure to draw an elliptical arc or a line with the **Arc** tool:

1. Click **Arc tool**.
2. Drag to draw an elliptical arc with the Arc tool.
3. Or, move the mouse in a horizontal or vertical direction to draw a horizontal or vertical line.

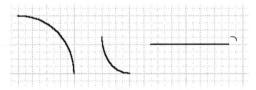

Drawing with the Rectangle Tool

Use the following procedure to draw a rectangle or a square with the **Rectangle** tool:

1. Click **Rectangle tool**.
2. Drag diagonally to draw a rectangle with the Rectangle tool. By default, the rectangle is filled with a white fill.
3. Or, hold down the **Shift** key to draw a square.

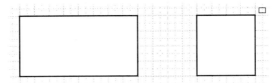

Drawing with the Ellipse Tool

Use the following procedure to draw an ellipse or a circle with the **Ellipse** tool:

1. Click **Ellipse tool**.

2. Drag diagonally to draw an ellipse with the Ellipse tool. By default, the ellipse is filled with a white fill.

3. Or, hold down the **Shift** key to draw a circle.

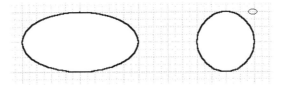

Hands-On Activity

In this activity, you use the drawing functions. Begin by starting Visio. Then open a new document with no template.

1. Click the **Pencil Tool** button. Notice that the cursor looks like a tiny pencil with a small crosshair at the "pencil" tip.

2. Drag across the page (hold down the button, move the mouse, release the button). Notice that Visio has drawn a line (or an arc, depending on your mouse movement).

 Point of Interest: While holding down the mouse button for the Pencil tool, carefully examine the cursor. It has changed to a + marker, along with either (1) a short straight line; or (2) a small arc. The straight line means Visio has sensed your mouse movement to be in a straight line; Visio accordingly draws a straight line.

The small arc means Visio has sensed your mouse movement to be curved; Visio accordingly draws an arc. If the arc looks to you like a straight line—even though Visio shows the arc indicator—it probably is a straight line.

3. Using the Pencil tool is tricky. Practicing drawing lines with the **Pencil** tool by moving the mouse straight.

4. Practice drawing arcs with the **Pencil** tool by moving the mouse in a curve. I find it easier to move the mouse slowly and in an exaggerated curving motion.

5. Hold down the **Shift** key while drawing with the Pencil tool. Notice how Visio constrains the line to 45 or 90 degrees and the arc's shape.

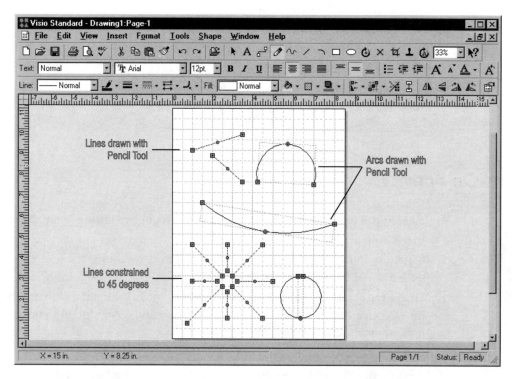

6. Select one of the arcs you have drawn.

🖎 **Point of Interest:** Notice the lines and arcs have three green squares. At one end, the green square has a small x in it. That handle is Visio's reminder of where you began drawing the line or arc. At the other end, the green square has a small + in it. That is where you ended drawing the line or arc. At the center, the third green square is empty. This handle shows you the center of the line or arc. It also functions as a *control point*. When you grab it and move it toward the midpoint of the other two handles, an arc straightens out into a line. Moving the midpoint of a line moves the line.

7. Drag the center handle back and forth. Notice how the arc changes its curve.

8. Press **Ctrl+A** to select all objects in the drawing.

9. Press **Delete** to erase all objects.

10. Select the **Rectangle** tool.

11. Drag to draw a rectangle. Notice the rectangle is filled with white.

12. Select gray from the **Fill** list box.

13. Hold the **Shift** key while dragging with the **Rectangle** tool to draw a square.

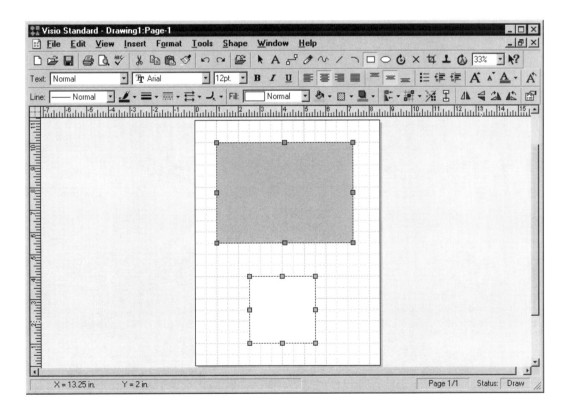

14. Press **Alt+F4** to exit Visio. Click **No** in response to the Save Changes dialog box.

This completes the hands-on activity for using the drawing tools. While many Visio drawings can be created by just using stencil shapes and connector tools, it is useful to know the drawing tools.

Module 24

Placing Text and Fields

Text (Insert | Field)

Uses

The **Text** tool, found on the toolbar, is used to place lines of text. Once the text is placed, you change the size, font, color, etc., as described in Module 14, Formatting Text.

A *field* is a piece of text attached to a shape that automatically updates itself. For example, the **Date/Time** field displays the current date and time, making it a time stamp. Some fields update themselves each time the drawing is opened; other fields update when the attached shape changes.

Visio lets you include special characters in the text. These characters are ones that you cannot normally type at the keyboard:

Special Character	Keystroke
' Beginning single-quote	Ctrl+[
' Ending single-quote	Ctrl+]
" Beginning double-quote	Ctrl+Shift+[
" Ending double-quote	Ctrl+Shift+]
• Bullet	Ctrl+Shift+8
– (en dash)	Ctrl+=
— (em dash)	Ctrl+Shift+=
- (discretionary hyphen)	Ctrl+hyphen
- (nonbreaking hyphen)	Ctrl+Shift+hyphen
/ (nonbreaking slash)	Ctrl+Shift+/
\ (nonbreaking backslash)	Ctrl+Shift+\
§ (section)	Ctrl+Shift+6
¶ (paragraph)	Ctrl+Shift+7

Special Character	Keystroke
© (copyright)	Ctrl+Shift+C
® (registered trademark)	Ctrl+Shift+R

A *discretionary hyphen* shows Visio where to hyphenate a word, if necessary, such as Micro-soft; the hyphen is normally invisible. A *nonbreaking hyphen* tells Visio not to hyphenate the word at the hyphen, such as "on-line." The *nonbreaking slash* is useful for fractions, such as "11/32."

Procedures

Before presenting the general procedures for placing text and fields, it is helpful to know about the shortcut key. These are:

Function	Keys	Menu	Toolbar Icon
Text Tool	Ctrl+2		A
Field	Ctrl+F9	Insert │ Field	

Text Field Functions	Keystroke
Height of text	Ctrl+Shift+H
Rotation angle of text	Ctrl+Shift+A
Width of text	Ctrl+Shift+W

Placing Text

Use the following procedure to add text to a page:

1. Select the **Text** tool from the toolbar.
2. Click on the page at the location you want the text.
3. Begin typing.
4. Select the **Pointer** tool when finished typing.

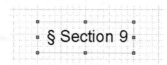

5. Resize and position the text block by dragging on the handles.

6. Format the text, as described in Module 14.

Inserting a Text Field

Use the following procedure to insert a text field:

1. Select a shape.
2. Select **Insert | Field**. Notice the Field dialog box.

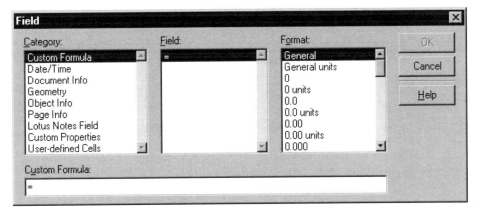

3. Select a **Category**:

   Custom Formula
   Date/Time
   Document Info
   Geometry
   Object Info
   Page Info
   Lotus Notes Field
   Custom Properties
   User-defined Cells

4. Select an item in the **Field** column.
5. Select an item in the **Format** column.
6. Click **OK**. Notice that Visio adds the text to the shape.

Hands-On Activity

In this activity, you use the text and field functions. Begin by starting Visio. Then open a new, blank document.

1. Select the **Text** tool.
2. Click anywhere on the page.
3. Type **This drawing was created on:**.

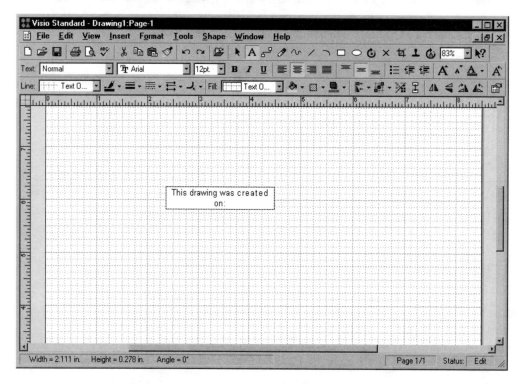

4. Select **Insert | Field**. Notice the Field dialog box.
5. Select **Date/Time** from the **Category** list.
6. Select **Creation Date** from the **Field** list.
7. Select **General (Long)** from the **Format** list.

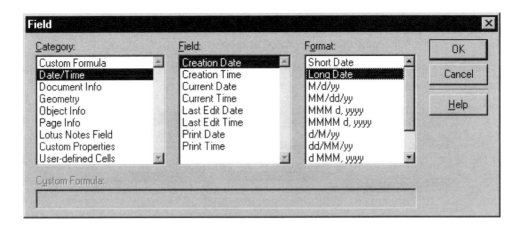

8. Click **OK**. Notice the date text placed in the drawing. (The date displayed in your drawing depends on the date you perform this task.)

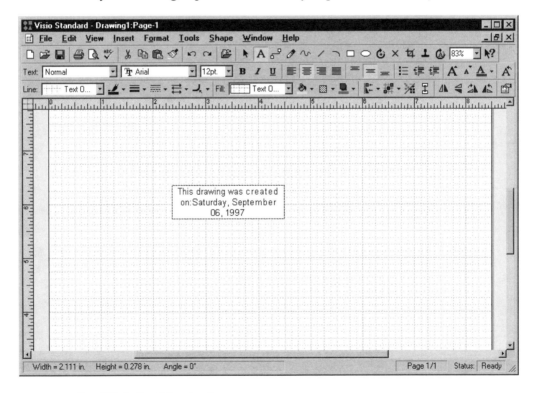

9. Click the **Line** tool.

10. Draw a line at any angle on the lower half of the page.

11. Select **Insert | Field**, ensuring that the line is still selected.
12. Select **Geometry** from the **Category** list.
13. Select **Angle** from the **Field** list.
14. Select **Degrees** from the **Format** list.

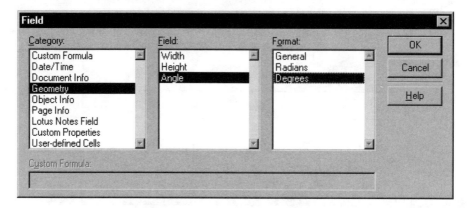

15. Click **OK** to dismiss the Field dialog box.
16. If necessary, zoom in to better read the test. Notice the text "27.65 deg." (The text displayed in your drawing may differ from the illustration below, depending on the angle that you drew the line.)

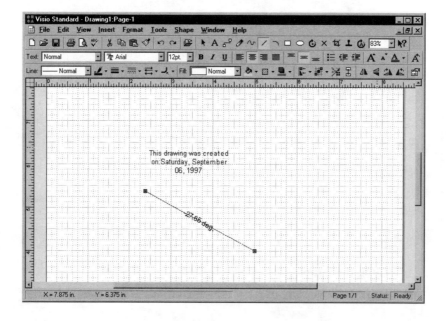

17. Click the **Pointer** tool.

18. Move an endpoint of the line to change its angle. Notice that the text automatically updates itself (to "56.84 deg." in my drawing).

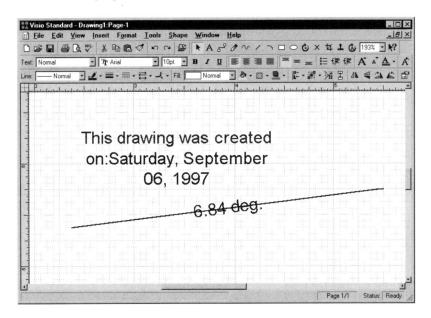

19. Press **Alt+F4** to exit Visio. Click **No** in response to the Save Changes dialog box.

This completes the hands-on activity for placing text and inserting fields. While text is important for annotating the drawing, the field feature is a very powerful way to display information that automatically updates itself.

Module 25

Spelling

Tools | Spelling

Uses

The **Spelling** function, found on the **Tools** menu, is used to check the spelling of words in the current drawing. Be careful: the speller does not check the spelling of words. Instead, it looks for words it does not recognize (ie, words it does not find in its lengthy word list). The speller has no way of knowing whether a correctly spelled word is used incorrectly, such using "weather" instead of "whether."

Procedures

Before presenting the general procedures for spelling, it is helpful to know about the shortcut keys. These are:

Function	Keys	Menu	Toolbar Icon
Spell Check	F7	Tools	Spelling	ABC✓

Check Spelling

Use the following procedure to check the spelling of all words in the current page:

1. Select **Tools | Spelling**.
2. When Visio finds a word it does not recognize, it displays the Spelling dialog box.

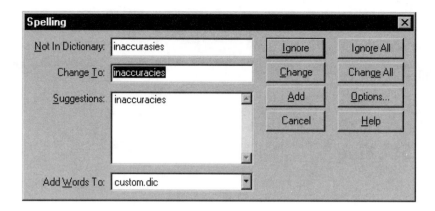

3. Click **Ignore** to have the Speller skip checking the word.

4. Click **Change** to select the work the Speller guesses might be correct.

5. Click **Add** to add the word to Speller's word list.

6. Click **Options** to display the Spelling Options dialog box.

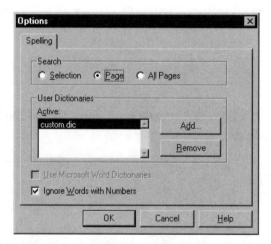

7. Use the **Search** section to check the selected text, all text on the current page, or all text on all pages.

8. Use **User Dictionaries** to maintain different custom dictionaries for different disciplines.

9. Use the **Use MS Word Dictionaries** to use the dictionary provided with Microsoft Word; if that dictionary is not installed, this option is grayed out.

10. Use **Ignore Word with Numbers** to prevent spell checking words that contain numbers such as dimensions and angles.

11. Click **OK**.

12. Click **OK** in the Spelling dialog box when the spell check is complete.

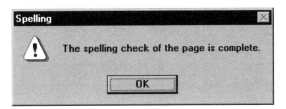

Hands-On Activity

In this activity, you use the spelling function. Begin by starting Visio. Then open a blank, new document.

1. Select the **Text** tool.

2. Type the following sentence:

   How to save Visio dramings.

   Notice the spelling error, "dramings."

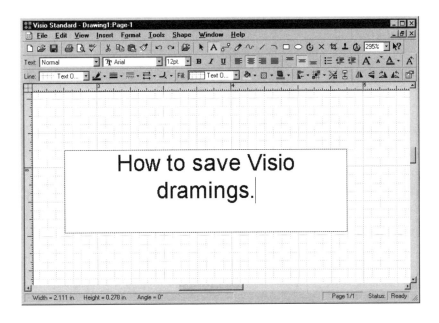

3. Select **Tools | Spelling** or press function key **F7**. Notice the Spelling dialog box shows "dramings" is not in its dictionary.

4. Click **Change**.

5. Visio displays a dialog box to indicate the spelling check is complete. Click **OK**. Notice that the spelling is correct in the drawing.

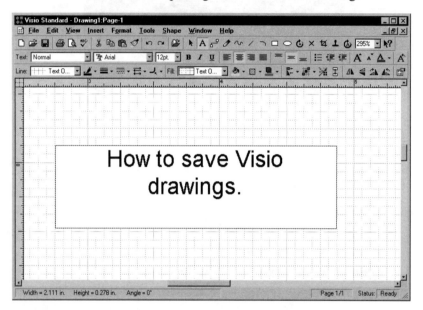

6. Exit Visio.

This completes the hands-on activity for spell checking text.

Module 26:

Finding and Replacing Text

Edit | Find (Replace)

Uses

The **Find** and **Replace** selections of the **Edit** menu are used to find and replace text in the current drawing. You can have Visio search just the selected text block, the entire page, or all pages of the current drawing.

Visio lets you search for some special characters that cannot be typed at the keyboard. These are:

Special Character	Meaning
^t	Tab
^r	Return (end of paragraph)
^-	Discretionary hyphen
^^	Caret (^) symbol
^?	Any character

Procedures

Before presenting the general procedures for finding and replacing text, it is helpful to know about the shortcut keys. These are:

Function	Keys	Menu
Find		Edit | Find
Replace		Edit | Replace
Repeat	F4	Edit | Repeat

Finding Text

Use the following procedure to find text:

1. Select **Edit | Find** to display the **Find** dialog.

2. Type the word or phrase to find in the **Find What** text box.
3. Click **Special** to insert a special character.
4. **Search**: find the phrase in:

   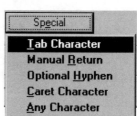

   ➤ **Selection**: the currently selected text.
   ➤ **Current Page**: the currently visible page (the default).
   ➤ **All Pages**: all pages of the drawing.
5. **Match Case**: the phrase must match the same pattern of upper and lowercase characters. When turned off (the default), the case is not matched. Searching for "visio" will find "Visio," "VISIO," and "visio."
6. **Find Whole Words Only**: the phrase must match the same set of words. When turned off (the default), the exact wording is not matched. Search for "visio" will find "Visio" and "vision."
7. Click **Find Next** to find the next occurrence of the word or phrase.

8. When you are done with finding words, click the **Cancel** button.

Replacing Text

Use the following procedure to replace text:

1. Select **Edit | Replace**.

2. Type the word or phrase to find in the **Find What** text box.
3. Type the replacement word or phrase in the **Replace With** text box.
4. Click **Special** to insert a special character.
5. Click **Replace** to replace the next occurrence of the word or phrase.
6. Or, click **Replace All** to replace all occurrences of the word.
7. When you are done replacing words, click the **Cancel** button.

Hands-On Activity

In this activity, you use the replace function. Begin by starting Visio.

1. Select **File | New | Browse Sample Drawings**.
2. From the **Business Diagram** folder, double-click on the **Global Organization Chart.Vsd** drawing. Notice that this is an 18-page drawing belonging to Cascadian Products, Inc.

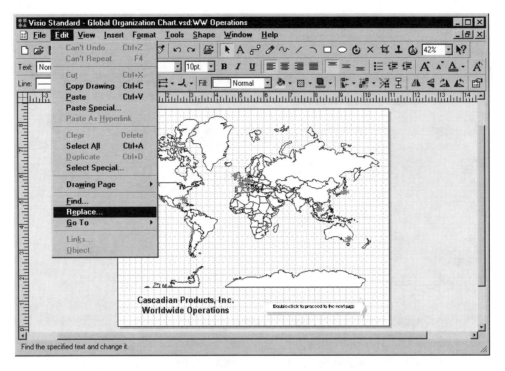

3. Select **Edit | Replace**. Notice the Replace dialog box.

4. Cascadian Products, Inc. has suffered a hostile corporate takeover. It is your job to implement the name change on all electronic documents. In the **Find What** text box, type **Cascadian**.

5. In the **Replace With** text box, type **Abbotsford**.

6. Click the **All Pages** button in the **Search** area to search all 18 pages of this drawing.

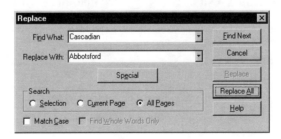

7. Click **Replace All**. Notice that Visio takes a few seconds to replace all occurrences of "Cascadian" with "Abbotsford." Notice the alert box when Visio is finished.

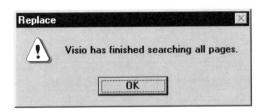

8. Click **OK**. Visio doesn't say so, but the name was changed on three of the eighteen pages.

9. Click **Cancel** to dismiss the Replace dialog box.

10. Select **View | Actual Size** to examine the changes to the text.

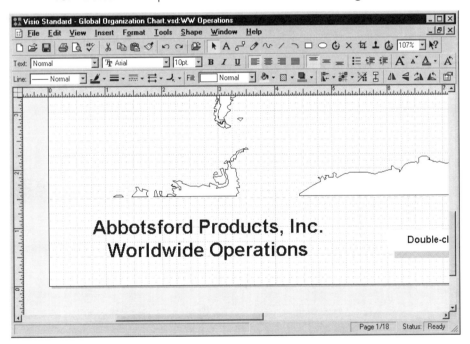

11. Press **Alt+F4** to exit Visio. Click **No** in response to the Save Changes dialog box.

This completes the hands-on activity for replacing text in a drawing.

Module 27

Dimensioning
(Visio Technical Only)

Uses

Architects and mechanical designers use dimensions to indicate the exact measurements of distances and angles. Specifying the distance is much more accurate than measuring off the paper with a ruler.

There are five basic types of dimensions: horizontal, vertical, aligned, angular, and diameter (which includes radius), as the illustration below shows.

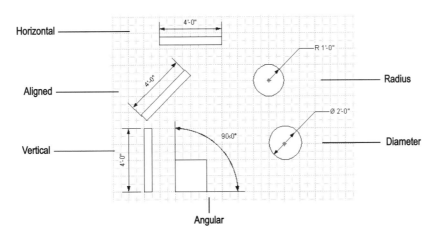

> **Horizontal dimension**: measures the horizontal distance between two points.
> **Vertical dimension**: measures the vertical distance between two points.
> **Aligned dimension**: measures the shortest distance between two points.
> **Angular dimension**: measures the angle between two lines.
> **Diameter dimension**: measures the radius or diameter of arcs and circles.

Other disciplines sometimes use special type of dimensions, known as ordinate dimensions and tolerance symbols.

The dimensioning functions are not found on any menu or toolbar. Instead, open the **General - Dimensions Architectural** or **Engineering** stencil file, found in the **Annotation** folder. The dimension stencils are provided only with Visio Technical. In Visio Standard and Professional, you create dimensions using these steps:

1. Draw a line with the **Line** tool.
2. Use **Format | Line** to add arrowheads at both ends.
3. Use **Insert | Field** to attach a **Geometry | Width | General Units** field to the line. (Substitute **Height** for vertical dimensions and **Angle** for angular dimensions.)
4. Use **Format | Text** to make the font a legible size.
5. Once the dimension line looks good to you, use **Format | Define Styles** to create text and line styles for the named dimension.

Procedures

There are no shortcut keys for dimensioning in Visio. Ensure the **Dimension Lines - Architectural** or **Mechanical** stencil is open.

Horizontal Dimensions

Use the following procedure to place a horizontal dimension:

1. Drag the **Horizontal** shape from the **Dimension Lines** stencil onto the page.
2. Drag one end of the horizontal dimension to one end of the shape being dimensioned.
3. Drag the other end of the horizontal dimension to the other end of the shape.
4. If required, adjust the location of the text and dimension height, as shown in the illustration.

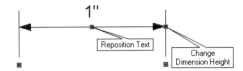

5. Adjusting the location of the text moves it away from the *dimension line*, the horizontal line.

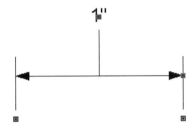

6. Changing the dimension height stretches the *extension lines*, the two vertical lines.

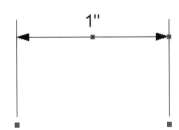

Vertical Dimensions

Use the following procedure to place a vertical dimension:

1. Drag the **Vertical** shape from the **Dimension Lines** stencil onto the page.

2. Drag one end of the vertical dimension to one end of the shape being dimensioned.

3. Drag the other end of the vertical dimension to the other end of the shape.

4. If required, adjust the location of the text and dimension height.

Aligned Dimensions

Use the following procedure to place an aligned dimension:

1. Drag the **Aligned Even** shape from the **Dimension Lines** stencil onto the page.

2. Drag one end of the aligned dimension to one end of the shape being dimensioned.

3. Drag the other end of the aligned dimension to the other end of the shape.

4. If required, adjust the location of the text and dimension height.

Angular Dimensions

Use the following procedure to place an angular dimension:

1. Drag the **Angle Even** shape from the **Dimension Lines** stencil onto the page.

2. Drag one end of the angle dimension to one end of the shape being dimensioned.

3. Drag the other end of the angle dimension to the other end of the shape.

4. If required, adjust the location of the text.

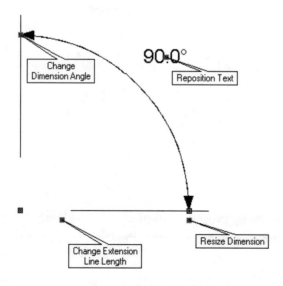

Radius Dimensions

Use the following procedure to place a radius dimension:

1. Drag the **Radius** shape from the **Dimension Lines** stencil onto the page.
2. Drag one end of the radius dimension to the center of the shape being dimensioned.
3. Drag the other end of the radius dimension to the edge of the shape.
4. If required, adjust the location of the text.

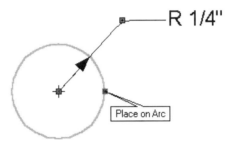

Diameter Dimensions

Use the following procedure to place a diameter dimension:

1. Drag the **Diameter** shape from the **Dimension Lines** stencil onto the page.
2. Drag one end of the diameter dimension to the center of the shape being dimensioned.
3. Drag the other end of the diameter dimension to the edge of the shape.

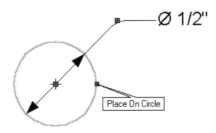

247

Hands-On Activity

In this activity, you use the dimensioning functions. Begin by starting Visio.

1. Open the drawing **Interior Wall Elevation.Vsd** file supplied in folder **Facilities Management**.

2. Open the stencil **General - Dimensioning Architecture.Vss** file found in folder **Annotation**.

3. Zoom into the wall elevation using the **Ctrl+Shift** windowing.

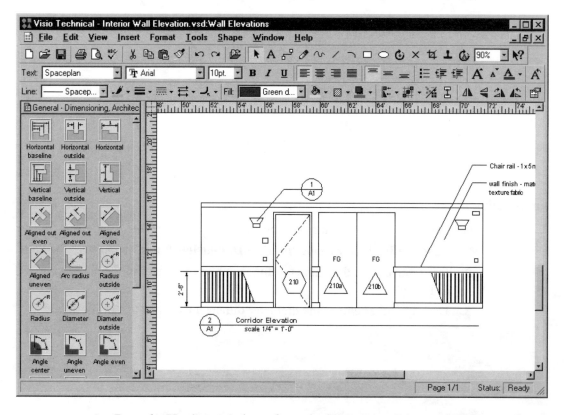

4. Drag the **Horizontal** shape from the **Dimension Lines - Architecture** stencil.

5. Ensure the left end of the **Horizontal** shape attaches to the upper left corner of the wall.

6. Drag the right handle of the **Horizontal** shape to the right end of the kitchen wall.

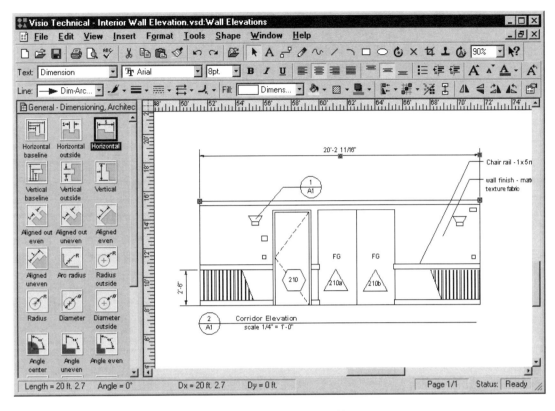

7. Drag the **Vertical** shape from the stencil.

8. Ensure the lower end of the **Vertical** shape attaches to the lower left corner of the left wall.

9. Drag the upper handle of the **Vertical** shape to the top end of the kitchen wall.

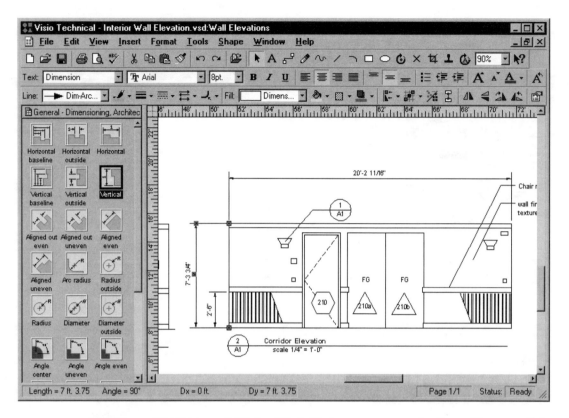

10. Press **Alt+F4** to exit Visio. Click **No** in response to the Save Changes dialog box.

This completes the hands-on activity for dimensioning.

Module 28

Inserting Objects

Insert | Object, Crop Tool

Uses

The insert functions, found on the **Insert** menu, are used to insert objects created by other Windows applications. From the **Insert** menu, Visio inserts the following kinds of objects: picture, control, clip art, Microsoft Graph, Microsoft WordArt, and object. In addition, Visio Technical inserts an AutoCAD file as an object (see Module 37).

In general, you can only view the inserted object. Visio allows you to perform some very limited editing functions on inserted objects. You can move and copy the object. While you cannot change the contents of the object, you can apply formatting to its frame, such as add a shadow and crop its boundary.

➤ **Insert | Picture**: Inserts a file in the following format created by the related applications:

Filename Extension	*Application File Format*
AI	Adobe Illustrator
BMP	Windows Bitmap
CDR	CorelDraw v3, 4, 5
CGM	Computer Graphics Metafile
CMX	Corel Clipart
DIB	Windows Bitmap
DRW	Micrografx Designer v3.1
DSF	Micrografx Designer v6
DWF	Drawing Web Format, Whip
EMF	Enhanced Metafile
EPS	Encapsulated PostScript
GIF	Graphic Interchange Format

Filename Extension	Application File Format
IGS	Initial Graphics Exchange Spec
JPG	JPEG Format
PCT	Mac Picture File Format
PCX	ZSoft PC-Paint Bitmap
PNG	Portable Network Graphics
PS	PostScript
TIF	Tagged Image File Format
WMF	Windows Metafile

 Point of Interest: BMP, DIB, and WMF are commonly used by the Windows applications when exchanging images via the Clipboard. TIFF, PCX, EPS, and PS are most commonly used for desktop publishing.

GIF and JPEG are the most commonly used file formats for display images on the Internet. Both file formats automatically compress the image to make its file size smaller; a smaller file transmits over the Internet faster than a larger file. However, because GIF translators require a licensing fee, the PNG format has been developed to replace GIF.

➤ **Insert | Control:** Inserts a VBA control. VBA is short for Visual Basic for Applications; see *Learn Visio 5.0 for the Advanced User* from Wordware Publishing, Inc.

➤ **Insert | AutoCAD Drawing**: AutoCAD v2.x through Release 13 (*Visio Technical only*). See Module 37.

➤ **Insert | Word Art**: Microsoft Word Art v2.0

➤ **Insert | ClipArt**: Microsoft Clip Art Gallery

➤ **Insert | Microsoft Graph**: Microsoft Graph v5.0

➤ **Insert | Object**: Any OLE-aware Windows application installed on your computer (this includes Paintbrush, PowerPoint, WordPad, Netscape Navigator, Excel, Media Clip, and AutoCAD).

Using one of the **Insert** commands on a "foreign" file differs from using the **Open** command:

Open Command	Insert Command
Translates the file into Visio format.	Keeps the object in its original format.
Slower.	Faster.

Open Command	*Insert Command*
Cannot be linked.	Can be linked back to source application.
Editable by Visio.	Not editable by Visio; can be edited by the source application and results seen in Visio.

When you insert an object, you can instruct Visio to maintain a link back to the original file. Whenever that file is updated, the image in Visio is also updated.

Procedures

Before presenting the general procedures for inserting objects, it is helpful to know about the menu keystrokes. These are:

Function	Menu	Toolbar
Insert picture	Insert | Picture	
Insert VBA controls	Insert | Controls	
Insert clip art	Insert | Clip Art	
Insert Word Art	Insert | WordArt	
Insert graph	Insert | Microsoft Graph	
Insert DWG drawing	Insert | AutoCAD Drawing	
Insert OLE object	Insert | Object	✄
Crop object		

✎ **Point of Interest:** Visio can insert objects by four methods: picture, icon, object, and linked object. However, the picture, object, and linked object all look identical in the Visio drawing. To find out how the object was inserted, right-click the object to display the menu. The last item on the menu is your clue:

Menu Wording	*Meaning*
-none-	Inserted as a picture.
Object	Inserted as an object.
Linked Object	Inserted as a linked object.

Inserting a Picture

Use the following procedure to insert a picture file into the Visio drawing:

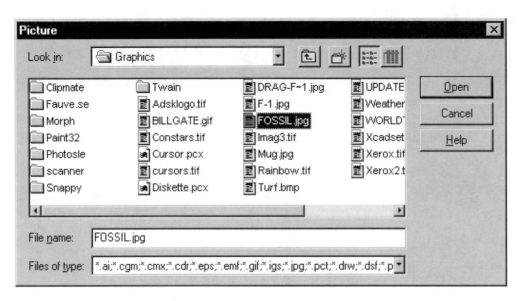

1. Select **Insert | Picture**. Notice that Visio displays the Picture dialog box.

2. Go to the drive and folder holding the picture you wish to insert.

3. Click **Open**.

4. Depending on the filetype selected, Visio may display an Import dialog box, which displays options specific to the filetype. For example, the illustration

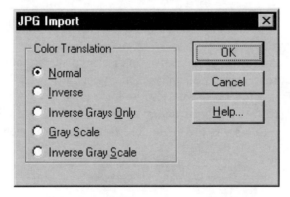

shows the JPG Import dialog box with color translation options:

5. If required, change the type of **Color Translation**:

Color Translation	Meaning
Normal	Keep colors the way they appear in the file.

Color Translation	Meaning
Inverse	Invert all colors, giving you a negative image.
Inverse Grays Only	Invert only black, white, and gray colors.
Gray Scale	Convert all colors to shades of gray.
Inverse Gray Scale	Convert all colors to shades of gray, then invert them.

The illustration shows the difference between normal (left) and inverse (right).

If required, change the **Retain Background** option. When turned on, Visio draws a background rectangle in the image's background color.

If required, set up the **Emulate Line Styles**. When turned on, Visio translates

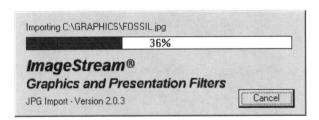

thick lines into polygons to help preserve visual accuracy.

6. Click **OK**. Notice that Visio takes several seconds to import the picture.

7. Notice that the picture object is placed in the center of the drawing, with green handles. You can move and resize the picture object just like a Visio shape.

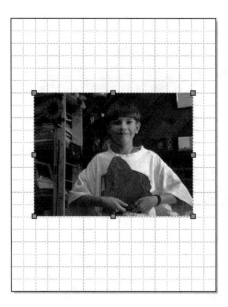

Applying the Crop Tool

Use the following procedure to crop a picture in the Visio drawing:

1. Select the picture object. Notice that Visio places handles around the picture.

2. Click the **Crop Tool** button.
3. Move the cursor over any of the handles. Notice that the cursor changes to a double-headed arrow
4. Press the mouse button and drag the handle inward to crop the picture.

5. Release the mouse button and Visio displays the smaller image. You can crop a picture object wider or narrower. When cropping wider, you don't see any more than what was present in the original image. However, the wide crop does affect the placement of fill and shadow.

6. While the Crop Tool is active, you may grab the center of the image and move it around. Notice that the cursor changes to an open hand; as you move (or *pan*) the image within its alignment box, the image turns black-white to improve the panning speed.

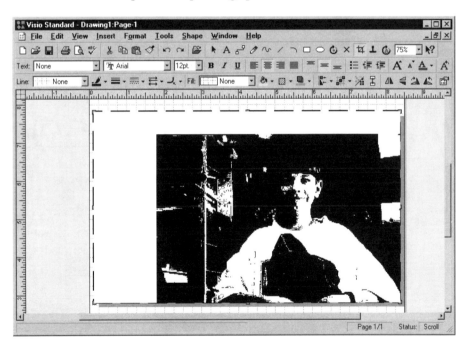

Insert a New Object

Use the following procedure to insert a new object:

1. Select **Insert | Object**. Notice that Visio displays the Insert Object dialog box.

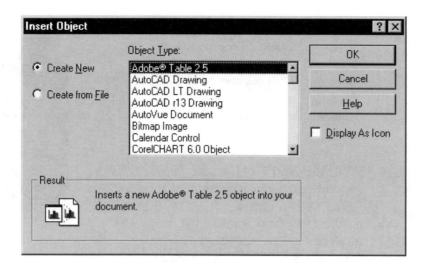

2. Click the **Create New** radio button. Notice the list of software names in the Object Type list box.

3. Select a software program name from the Object Type list box.

4. If required, click **Display As Icon**. Notice that the dialog box displays the icon associated with the selected software program.

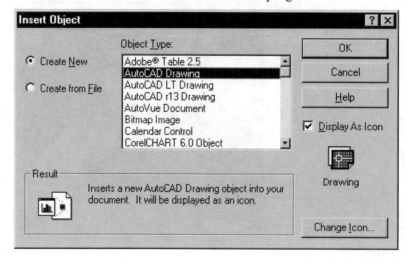

Display As Icon	*Meaning*
On	Object is displayed as an icon. An icon displays faster than the object. You cannot see the contents until you double-click the icon.
Off	Object is displayed as itself. The objects display slower than the icon representation. You see the contents of the object.

The illustration shows how the object looks when it is inserted as an icon in the Visio drawing.

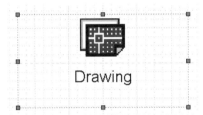

To change the icon, click the **Change Icon** button. The dialog box lets you select a different icon and type the wording you wish to appear under the icon (in the Label text box.)

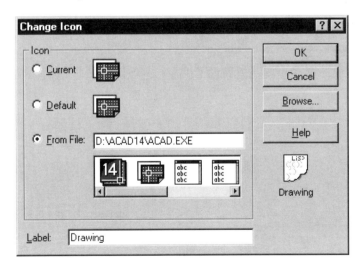

5.   Click **OK**. Notice that Windows launches the application.

6.   Create the new document within the application.

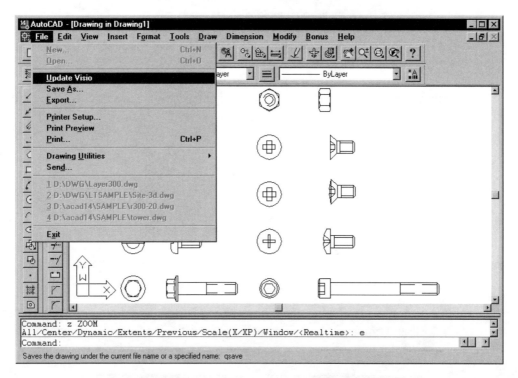

7.   Select **File | Exit & Return to filename.vsd** when you are ready to return to Visio with the object.

8.   Notice the object is placed in the center of the Visio page.

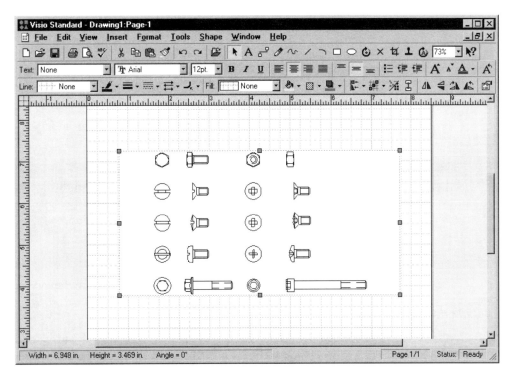

Inserting an Object From a File

Use the following procedure to insert an object from an existing file:

1. Select **Insert | Object**. Notice that Visio displays the Insert Object dialog box.

2. Click the **Create From File** radio button. Notice the File text box and the Browse button.

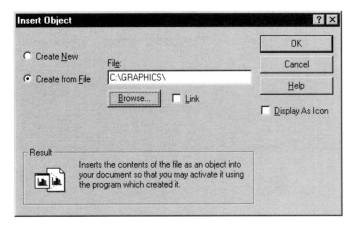

3. Type the name of the file; if you do not know it, click the **Browse** button to find it.

4. If required, click **Display As Icon**.

5. Click **OK**. Notice that Windows launches the application (called the *source* application) with the file you specified.

6. At the same time, the file appears in Visio, centered in the drawing.

 Point of Interest: In some cases, the file may appear in Visio as an icon when you don't specify that. This means that the source application is not fully operable with Visio. Instead, Windows places a *package* object, which looks like the icon. Right-click the icon to edit the package back in its source application.

7. You can exit the source application. Notice the object remains in Visio.

Placing a Linked Object

One problem with placing an object in Visio is that it may not be up to date. For this reason, Visio allows you to insert a *linked* object. Whenever the source application makes a change to the original file, the object in Visio is also updated. Use the following procedure to insert a linked object:

1. Select **Insert | Object**. Notice that Visio displays the Insert Object dialog box.

2. Click the **Create From File** radio button. Notice the File text box and the Browse button.

3. Type the name of the file; if you do not know it, click the **Browse** button to find it.

4. Click the **Link** check box.

5. Click **OK**. Notice that Windows launches the application (called the *source* application) with the file you specified.

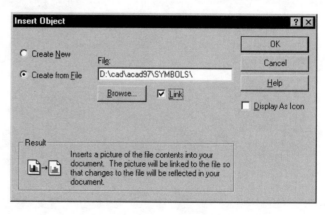

6. At the same time, the file appears in Visio, centered in the drawing.

7. A linked object looks no different from an unlinked object. To check if the object is linked, right-click on it. Notice the words "Linked Object" on the menu that appears. Although the menu may read Convert Link Drawing Object, Visio cannot convert drawing objects to Visio shapes.

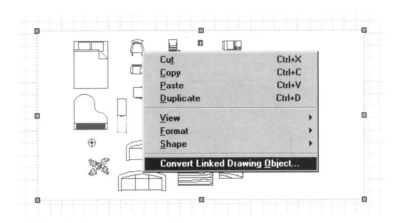

8. You can exit the source application. Notice the object remains in Visio.

Editing an Object

Use the following procedure to edit an object inserted in the Visio drawing:

1. Select the object.

2. Right-click the OLE object. Notice the menu that appears has three options for Object: Edit, Open, and Convert.

> **Edit** opens the source application with the object. When finished editing in the source application, select **File | Update** from the menu bar.

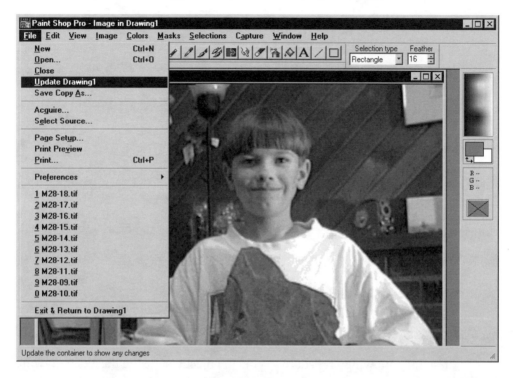

> **Open**: opens the source application with the object. When finished viewing in the source application, select **File | Update** from the menu bar.

> **Convert**: displays a dialog box listing the formats the object can be converted into. In most cases, the object *cannot* be converted to Visio shapes.

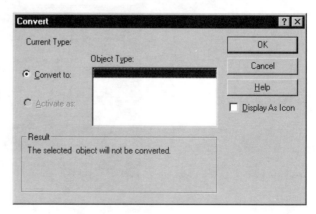

3. Press **Del** to erase the selected object from the drawing.

Updating a Linked Ojbect

Use the following procedure to update a linked object:

1. Select **Edit | Links** from the menu bar. Notice the Links dialog box, which lists all linked objects Visio found in the drawing.

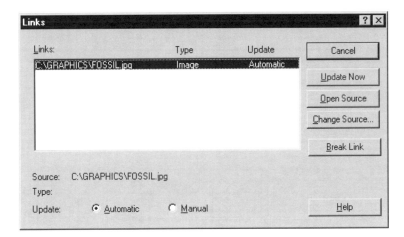

2. Notice that **Links** lists the names of linked files. **Type** describes the object. **Update** reports whether the object is automatically or manually updated.

3. To change the type of updating, click **Automatic** or **Manual**. *Automatic* updating means the object in Visio is updated whenever the source file changes; *manual* updating means you must use this dialog box and click the **Update Now** button. In practice, I find that automatic updating does not usually work.

4. **Open Source** opens the source application, along with a copy of the file. **Change Source** lets you specify a different file, such as a different image file; it does *not* change the source application.

5. **Break Link** breaks the link back to the source file. The object becomes unlinked.

6. When done, click **Close**.

265

Module 29

Exporting Drawings

File | Save As

Uses

The export function, found on the **Save As** dialog box from the **File** menu, converts the Visio drawing into other file formats. Visio drawings need to be converted since no other software program reads Visio files. You can export the current Visio page or selected shapes; you cannot export more than one page at a time. You select from the following file formats:

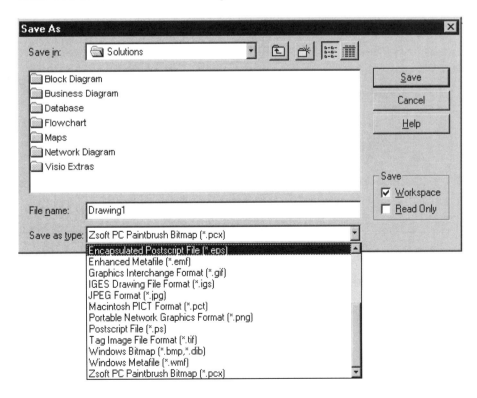

Program or File Format	File Type	Notes
Adobe Illustrator	AI	Vector; also saves in EPS format.
AutoCAD binary(*)	DWG	Vector; Release 10 through 13.
AutoCAD ASCII(*)	DWG	Vector; Release 10 through 13.
Computer Graphics Metafile	CGM	Vector-raster; saves in many flavors of CGM.
Drawing Web Format(*)	DWF	Vector; meant for use on the Internet.
Encapsulated PostScript	EPS	Vector; also saves in AI format.
Enhanced Metafile	EMF	Vector-raster.
Graphics Interchange Format	GIF	Raster; commonly used on the Internet.
HTML	HTM	Hypertext Markup Language.
Initial Graphics Exchange Spec	IGS	Vector; read by some CAD systems.
JPEG	JPG	Raster; commonly used on the Internet.
Macintosh Picture Format	PCT	Vector-raster.
PC-Paintbrush	PCX	Raster; commonly used for desktop publishing.
Portable Network Graphics	PNG	Raster; designed to replace GIF.
PostScript	PS	Vector; commonly used by desktop publishing.
Tagged Image File Format	TIFF	Raster; commonly used for desktop publishing.
Visio 4.0	VSD	Read by earlier versions of Visio.
Windows Bitmap	BMP	Raster; used by the Windows Clipboard.
Windows Metafile	WMF	Vector-raster; no options.

(*) Available only in Visio Technical.

Exporting in DWF and HTML formats are described in Module 35. Exporting in AutoCAD DWG and DXF file formats is described in Module 37.

 Point of Interest: When you have a choice of file formats and options, in general it is better to:

➤ Use a vector format since Visio is a vector format. Raster loses "resolution," which is a measure of accuracy.

➤ Use *compression* to reduce the file size of raster formats. Compression reduces the size of the file without losing any data.

➤ Not use compression when the receiving software cannot read it or when the format does not have the option.

Procedures

Before presenting the general procedures for exporting Visio files, it is helpful to know about the shortcut key. It is:

Function	Keys	Menu
Export	F12	File | SaveAs

Export to AI

Use the following procedure to convert the Visio drawing to Adobe Illustrator format:

1. Ensure a drawing is open in Visio. (Visio will export an empty drawing but there isn't much point to that exercise.)

2. Select **File | Save As**. Notice the Save As dialog box.

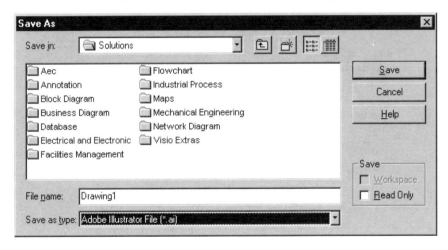

3. Click on the **Save as type** list box.

4. Select **Adobe Illustrator (*.ai)** file format.

5. If necessary, change the filename and select another folder or drive.

6. Click **Save**. Notice the AI/EPS Output Filter Setup dialog box.

7. Select **Standard Options -AI** from the **Profiles** list box.

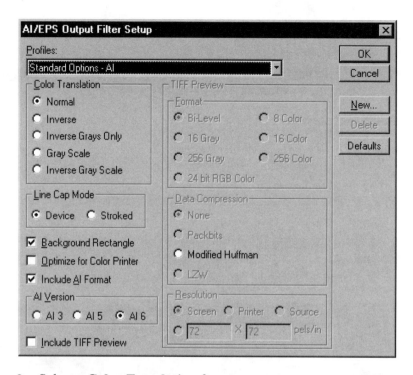

8. Select a **Color Translation** format:
    ➤ **Normal**: Keep the colors as they are in the Visio drawing.
    ➤ **Inverse**: Invert the colors (the picture looks like the strip of negatives from color film).
    ➤ **Inverse Grays Only**: Invert only white, black, and grays; keep other colors as they are.
    ➤ **Gray Scale**: Change the colors to levels of gray.
    ➤ **Inverse Gray Scale**: Change the colors to levels of gray and make a negative.

9. Select an option from **Line Cap Mode**:
    ➤ **Device**: Use the line style and width capabilities of the display and printer driver.
    ➤ **Stroked**: Thick and patterned lines are drawn as polygons to more closely represent how they appear in the Visio drawing.

10. Select from the following options:
    ➤ **Background Rectangle**: Places a rectangle around the drawing as a border.
    ➤ **Optimize for Color Printer**: Optimized export format for color printers.

➤ **Include AI Format**: Include data for Adobe Illustrator format.

➤ **AI Version**: Select from version 3, 5, or 6.

➤ **Include TIFF Preview**: Includes a small raster image for use within desktop publishing software, which cannot display AI and EPS formats.

➤ **Modified Huffman**: Compresses the data so that the file size is smaller.

11. When you make changes to the options and wish to save the changes for the next time you use this command, click **New**. Notice the New Profile Menu dialog box.

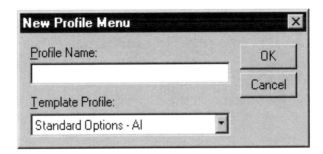

12. Type a name in the **Profile Name** text box. This name will appear in the Profiles list box the next time you see the AI/EPS Output Filter Setup dialog box.

13. Click **OK** to dismiss the dialog box.

14. Click **OK**. Visio converts the drawing and saves it in AI format.

Export to BMP or DIB

Use the following procedure to convert the Visio drawing to BMP (short for bitmap) or DIB (short for device independent bitmap) file format:

1. Ensure a drawing is open in Visio.

2. Select **File | Save As**.

3. Click on the **Save as type** list box and select **BMP** file format.

4. Click **Save**. Notice the BMP Output Filter Setup dialog box.

5. Select a **Profile**. Visio comes with two profiles: standard options and Visio HTML export.

6. Select **Format** options:

    ➤ **Bi-Level**: Converts Visio drawing colors to black or white.

    ➤ **16 Colors**: Reduces Visio drawing colors to 16 colors of the Windows standard.

    ➤ **16 Color Palette**: Reduces Visio drawing colors to 16 colors to the best approximation of the original.

    ➤ **256 Color**: Retain all Visio drawing colors.

7. Select a **Resolution** for the preview image:

    ➤ **Screen**: Use the screen resolution, typically 72dpi (dots per inch) or 96dpi.

    ➤ **Printer**: Use the printer's resolution, typically 300dpi or 600dpi; use this for the best quality hardcopy.

    ➤ **Source**: Let the destination application figure out the best resolution to use.

    ➤ **Custom**: Specify any resolution; default = 72x72dpi. Visio recommends that the resolution range between 32dpi and 400dpi.

8. Select a **Size** of image.

9. Select a **File Type**. Normally, you would keep **Win 3.X** but if you plan to use the BMP file with the OS/2 operating system, then select the **OS2 PM 1.X** option.

10. Select a **Color Translation** format:

    ➤ **Normal**: Keep the colors as they are in the Visio drawing.

    ➤ **Inverse**: Invert the colors (the picture looks like the strip of negatives from color film).

    ➤ **Inverse Grays Only**: Invert only white, black, and grays; keep other colors as they are.

    ➤ **Gray Scale**: Change the colors to levels of gray.

    ➤ **Inverse Gray Scale**: Change the colors to levels of gray and make a negative.

11. Select a method of **Data Compression**. While the file size is greatly reduced using **RLE** (short for "run length encoding"), most applications cannot read compressed BMP files. Therefore, leave the choice at **None**.

12. When you make changes to the options and wish to save the changes for the next time you use this command, click **New**.

13. Type a name in the **Profile Name** text box. This name will appear in the Profiles list box the next time you see the PNG Output Filter Setup dialog box.

14. Click **OK** to dismiss the dialog box.

15. Click **OK**. Visio converts the drawing and saves it in PCX format.

Export to CGM

Use the following procedure to convert the Visio drawing to computer graphics metafile format:

1. Ensure a drawing is open in Visio.

2. Select **File | Save As**.

3. Click on the **Save as type** list box and select **CGM** file format.

4. Click **Save**. Notice the CGM Output Filter Setup dialog box.

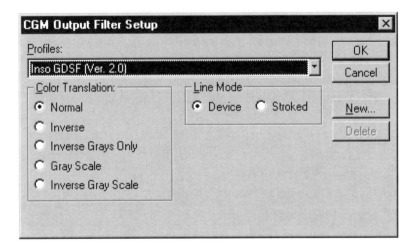

5. Select a **Profile** from one of the following CGM flavors. Check with your software documentation to see which one might work best:
   - ➤ Standard Options (ANSI)
   - ➤ Mil-D-28003 (CALS)
   - ➤ Mil-D-28003A (CALS)
   - ➤ ANSI CGM 3.0
   - ➤ INSO GDSF (v1.0)
   - ➤ INSO GDSF (v2.0)

6. Select a **Color Translation** format:
   - ➤ **Normal**: Keep the colors as they are in the Visio drawing.

273

> ➤ **Inverse**: Invert the colors (the picture looks like the strip of negatives from color film).

> ➤ **Inverse Grays Only**: Invert only white, black, and grays; keep other colors as they are.

> ➤ **Gray Scale**: Change the colors to levels of gray.

> ➤ **Inverse Gray Scale**: Change the colors to levels of gray and make a negative.

7. Select an option from **Line Mode**:

> ➤ **Device**: Use the line style and width capabilities of the display and printer driver.

> ➤ **Stroked**: Thick and patterned lines are drawn as polygons to more closely represent how they appear in the Visio drawing.

8. When you make changes to the options and wish to save the changes for the next time you use this command, click **New**.

9. Type a name in the **Profile Name** text box. This name will appear in the Pro-files list box the next time you see the CGM Output Filter Setup dialog box.

10. Click **OK** to dismiss the dialog box.

11. Click **OK**. Visio converts the drawing and saves it in CGM format.

Export to EPS and PS

Use the following procedure to convert the Visio drawing to PostScript and en-capsulated PostScript format:

1. Ensure a drawing is open in Visio.

2. Select **File | Save As**.

3. Click on the **Save as type** list box and select **EPS** or **PS** file format.

4. Click **Save**. Notice the AI/EPS Output Filter Setup dialog box.

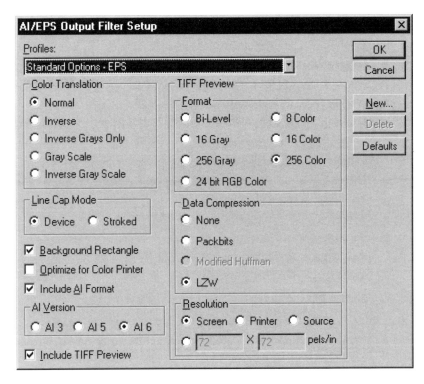

5. Select **Standard Options -EPS** from the **Profiles** list box.

6. Select a **Color Translation** format:
   ➤ **Normal**: Keep the colors as they are in the Visio drawing.
   ➤ **Inverse**: Invert the colors (the picture looks like the strip of negatives from color film).
   ➤ **Inverse Grays Only**: Invert only white, black, and grays; keep other colors as they are.
   ➤ **Gray Scale**: Change the colors to levels of gray.
   ➤ **Inverse Gray Scale**: Change the colors to levels of gray and make a negative.

7. Select an option from **Line Cap Mode**:
   ➤ **Device**: Use the line style and width capabilities of the display and printer driver.
   ➤ **Stroked**: Thick and patterned lines are drawn as polygons to more closely represent how they appear in the Visio drawing.

8. Select from the following options:

➤ **Background Rectangle**: Places a rectangle around the drawing as a border.

➤ **Optimize for Color Printer**: Optimized export format for color printers.

➤ **Include AI Format**: Include data for Adobe Illustrator format.

➤ **AI Version**: Select from version 3, 5, or 6.

➤ **Include TIFF Preview**: Includes a small raster image for use within desktop publishing software, which cannot display AI and EPS formats.

9. Select one of the **TIFF Preview** formats:

➤ **Bilevel**: Reduce the image to black and white; creates a smaller file size.

➤ **16 Gray**: sixteen shades of gray.

➤ **256 Gray**: 256 shades of gray.

➤ **8 Color**: eight colors.

➤ **16 Color**: sixteen colors.

➤ **256 Color**: 256 colors; the default.

➤ **24 Bit RGB Color**: 16.7 million colors.

10. Select a **Data Compression** for the preview image:

➤ **None**: no compression creates a larger file but may be necessary for some applications that cannot read compressed TIFF files.

➤ **Packbits**: best suited to black-white images.

➤ **Modified Huffman**: best suited for gray images.

➤ **LZW**: best suited for color images.

11. Select a **Resolution** for the preview image:

➤ **Screen**: Use the screen resolution, typically 72dpi (dots per inch) or 96dpi.

➤ **Printer**: Use the printer's resolution, typically 300dpi or 600dpi; use this for the best quality hardcopy.

➤ **Source**: Let the destination application figure out the best resolution to use.

➤ **Custom**: Specify any resolution; default = 72x72dpi. Visio recommends that the resolution range between 32dpi and 400dpi.

12. When you make changes to the options and wish to save the changes for the next time you use this command, click **New**.

13. Type a name in the **Profile Name** text box. This name will appear in the Profiles list box the next time you see the AI/EPS Output Filter Setup dialog box.

14. Click **OK** to dismiss the dialog box.

15. Click **OK**. Visio converts the drawing and saves it in EPS or PS format.

Export to EMF

Use the following procedure to convert the Visio drawing to enhanced Metafile format:

1. Ensure a drawing is open in Visio.

2. Select **File | Save As**.

3. Click on the **Save as type** list box and select **EMF** file format.

4. Click **Save**. Visio converts the drawing and saves it in EMF format. There are no options.

Export to GIF

Use the following procedure to convert the Visio drawing to graphics interchange format:

1. Ensure a drawing is open in Visio.

2. Select **File | Save As**.

3. Click on the **Save as type** list box and select **GIF** file format.

4. Click **Save**. Notice the GIF Output Filter Setup dialog box.

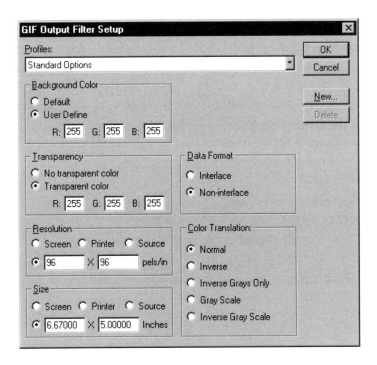

5. Choose a **Profile** from the list box. Visio includes two: Standard Options and Visio HTML Export. The latter is used in conjunction with exporting the drawing in HTML format for use in the Internet. See Module 35.

6. Choose a color for **Transparency**. *Transparency* allows other images or text "underneath" the GIF image to show through the color selected as transparent.

7. Select a **Resolution** for the preview image:

   ➤ **Screen**: Use the screen resolution, typically 72dpi (dots per inch) or 96dpi.

   ➤ **Printer**: Use the printer's resolution, typically 300dpi or 600dpi; use this for the best quality hardcopy.

   ➤ **Source**: Let the destination application figure out the best resolution to use.

   ➤ **Custom**: Specify any resolution; default = 72x72dpi. Visio recommends that the resolution range between 32dpi and 400dpi.

8. Select a **Size** of image.

9. Select a **Data Format**. *Interlaced* is useful for the Internet, since it shows parts of the image in the Web browser before the entire image has been delivered over the (relatively slow) telephone lines. *Non-interlaced* means the entire image must be delivered before it can be displayed.

10. Select a **Color Translation** format:

    ➤ **Normal**: Keep the colors as they are in the Visio drawing.

    ➤ **Inverse**: Invert the colors (the picture looks like the strip of negatives from color film).

    ➤ **Inverse Grays Only**: Invert only white, black, and grays; keep other colors as they are.

    ➤ **Gray Scale**: Change the colors to levels of gray.

    ➤ **Inverse Gray Scale**: Change the colors to levels of gray and make a negative.

11. When you make changes to the options and wish to save the changes for the next time you use this command, click **New**.

12. Type a name in the **Profile Name** text box. This name will appear in the Profiles list box the next time you see the GIF Output Filter Setup dialog box.

13. Click **OK** to dismiss the dialog box.

14. Click **OK**. Visio converts the drawing and saves it in GIF format.

Export to IGES

IGES is a file format commonly used by high-end CAD systems to exchange drawings. Typically, these CAD packages must translate the IGES file to their own format. Use the following procedure to convert the Visio drawing to initial graphics exchange specification:

1. Ensure a drawing is open in Visio.
2. Select **File | Save As**.
3. Click on the **Save as type** list box and select **IGS** file format.
4. Click **Save**. Notice the IGES Output Filter Setup dialog box.

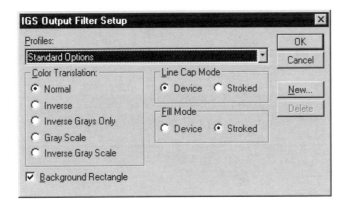

5. Select a **Color Translation** format:
   - ➤ **Normal**: Keep the colors as they are in the Visio drawing.
   - ➤ **Inverse**: Invert the colors (the picture looks like the strip of negatives from color film).
   - ➤ **Inverse Grays Only**: Invert only white, black, and grays; keep other colors as they are.
   - ➤ **Gray Scale**: Change the colors to levels of gray.
   - ➤ **Inverse Gray Scale**: Change the colors to levels of gray and make a negative.
6. **Background Rectangle** draws a rectangle around the extents of the drawing.
7. Select an option from **Line Cap Mode**:
   - ➤ **Device**: Use the line style and width capabilities of the display and printer driver.
   - ➤ **Stroked**: Thick and patterned lines are drawn as polygons to more closely represent how they appear in the Visio drawing.

8. Select an option from **Fill Mode**:

> ➤ **Device**: Use the fill capabilities of the display and printer driver.

> ➤ **Stroked**: Filled areas are drawn as polygons to more closely represent how they appear in the Visio drawing.

9. When you make changes to the options and wish to save the changes for the next time you use this command, click **New**.

10. Type a name in the **Profile Name** text box. This name will appear in the Profiles list box the next time you see the AI/EPS Output Filter Setup dialog box.

11. Click **OK** to dismiss the dialog box.

12. Click **OK**. Visio converts the drawing and saves it in IGES format.

Export to JPEG

JPEG is popular for very large images because it does a terrific job of compressing files. JPEG can make the file size smaller than any other file format; the drawback is that it is a *lossy* compression, which means some of the image may be distorted due to the compression process. "JPEG" is short for Joint Photographic Expert Group. Use the following procedure to convert the Visio drawing to JPEG format:

1. Ensure a drawing is open in Visio.

2. Select **File | Save As**.

3. Click on the **Save as type** list box and select **JPG** file format.

4. Click **Save**. Notice the JPG Output Filter Setup dialog box.

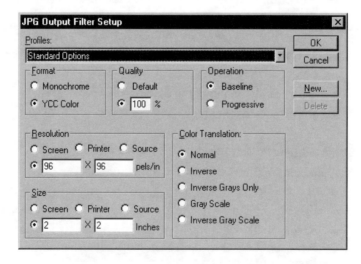

5. Choose a **Profile** from the list box. Visio includes two: Standard Options and Visio HTML Export. The latter is used in conjunction with exporting the drawing in HTML format for use in the Internet. See Module 35.

6. Choose a **Format** for monochrome or YCC color.

7. Choose a **Quality**, which is the JPEG term for "compression." The higher the quality (closer to 100%), the lower the compression and the larger the file size.

8. Choose an **Operation** for baseline of progressive.

9. Select a **Resolution** for the preview image:
   ➤ **Screen**: Use the screen resolution, typically 72dpi (dots per inch) or 96dpi.
   ➤ **Printer**: Use the printer's resolution, typically 300dpi or 600dpi; use this for the best quality hardcopy.
   ➤ **Source**: Let the destination application figure out the best resolution to use.
   ➤ **Custom**: Specify any resolution; default = 72x72dpi. Visio recommends that the resolution range between 32dpi and 400dpi.

10. Select a **Size** of image.

11. Select a **Color Translation** format:
    ➤ **Normal**: Keep the colors as they are in the Visio drawing.
    ➤ **Inverse**: Invert the colors (the picture looks like the strip of negatives from color film).
    ➤ **Inverse Grays Only**: Invert only white, black, and grays; keep other colors as they are.
    ➤ **Gray Scale**: Change the colors to levels of gray.
    ➤ **Inverse Gray Scale**: Change the colors to levels of gray and make a negative.

12. When you make changes to the options and wish to save the changes for the next time you use this command, click **New**.

13. Type a name in the **Profile Name** text box. This name will appear in the Profiles list box the next time you see the JPG Output Filter Setup dialog box.

14. Click **OK** to dismiss the dialog box.

15. Click **OK**. Visio converts the drawing and saves it in JPEG format.

Export to PICT

Use the following procedure to convert the Visio drawing to Macintosh picture file format:

1. Ensure a drawing is open in Visio.
2. Select **File | Save As**.
3. Click on the **Save as type** list box and select **PCT** file format.
4. Click **Save**. Notice the PICT Output Filter Setup dialog box.

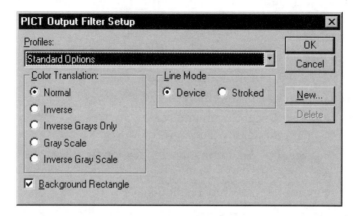

5. Select a **Color Translation** format:
   - ➤ **Normal**: Keep the colors as they are in the Visio drawing.
   - ➤ **Inverse**: Invert the colors (the picture looks like the strip of negatives from color film).
   - ➤ **Inverse Grays Only**: Invert only white, black, and grays; keep other colors as they are.
   - ➤ **Gray Scale**: Change the colors to levels of gray.
   - ➤ **Inverse Gray Scale**: Change the colors to levels of gray and make a negative.
6. **Background Rectangle** draws a rectangle around the extents of the drawing.
7. Select an option from **Line Cap Mode**:
   - ➤ **Device**: Use the line style and width capabilities of the display and printer driver.
   - ➤ **Stroked**: Thick and patterned lines are drawn as polygons to more closely represent how they appear in the Visio drawing.
8. When you make changes to the options and wish to save the changes for the next time you use this command, click **New**.
9. Type a name in the **Profile Name** text box. This name will appear in the Profiles list box the next time you see the AI/EPS Output Filter Setup dialog box.
10. Click **OK** to dismiss the dialog box.

11. Click **OK**. Visio converts the drawing and saves it in PICT format.

Export to PCX

Use the following procedure to convert the Visio drawing to PC-Paintbrush file format:

1. Ensure a drawing is open in Visio.
2. Select **File | Save As**.
3. Click on the **Save as type** list box and select **PCX** file format.
4. Click **Save**. Notice the **PCX Output Filter Setup** dialog box.
5. Select a **Profile**. Visio comes with three profiles: standard options, PC Paint-brush, and Visio HTML export.

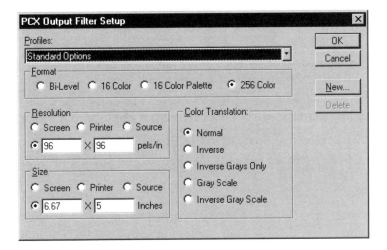

6. Select **Format** options:
   ➤ **Bi-Level**: Converts Visio drawing colors to black or white.
   ➤ **16 Colors**: Reduces Visio drawing colors to 16 colors of the Windows standard.
   ➤ **16 Color Palette**: Reduces Visio drawing colors to 16 colors to the best approximation of the original.
   ➤ **256 Color**: Retain all Visio drawing colors.
7. Select a **Resolution** for the preview image:
   ➤ **Screen**: Use the screen resolution, typically 72dpi (dots per inch) or 96dpi.
   ➤ **Printer**: Use the printer's resolution, typically 300dpi or 600dpi; use this for the best quality hardcopy.

283

> ➤ **Source**: Let the destination application figure out the best resolution to use.

> ➤ **Custom**: Specify any resolution; default = 72x72dpi. Visio recommends that the resolution range between 32dpi and 400dpi.

8. Select a **Size** of image.

9. Select a **Color Translation** format:

> ➤ **Normal**: Keep the colors as they are in the Visio drawing.

> ➤ **Inverse**: Invert the colors (the picture looks like the strip of negatives from color film).

> ➤ **Inverse Grays Only**: Invert only white, black, and grays; keep other colors as they are.

> ➤ **Gray Scale**: Change the colors to levels of gray.

> ➤ **Inverse Gray Scale**: Change the colors to levels of gray and make a negative.

10. When you make changes to the options and wish to save the changes for the next time you use this command, click **New**.

11. Type a name in the **Profile Name** text box. This name will appear in the **Profiles** list box the next time you see the **PNG Output Filter Setup** dialog box.

12. Click **OK** to dismiss the dialog box.

13. Click **OK**. Visio converts the drawing and saves it in PCX format.

Export to PNG

PNG was invented to replace GIF, which requires royalty payments under certain situations. Use the following procedure to convert the Visio drawing to portable network graphics format:

1. Ensure a drawing is open in Visio.

2. Select **File | Save As**.

3. Click on the **Save as type** list box and select **PNG** file format.

4. Click **Save**. Notice the PNG Output Filter Setup dialog box.

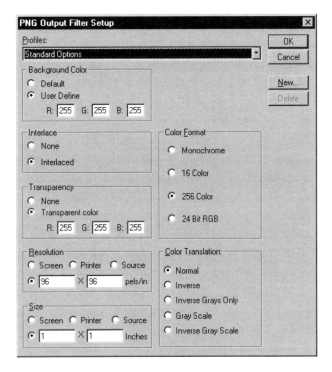

5. Choose a **Profile** from the list box. Visio includes two: Standard Options and Visio HTML Export. The latter is used in conjunction with exporting the drawing in HTML format for use in the Internet. See Module 35.

6. Select a **Background Color** of white (the default) or specify another color using the RGB (red, green, blue) method:

RGB	Meaning
0 0 0	Black
255 0 0	Red
0 255 0	Green
0 0 255	Blue
191 191 191	Medium Gray
255 255 255	White

The RBG method allows you to specify 16.7 million colors by varying the amount of red, green, and blue in 256 increments.

7. Decide whether you want to **Interlace**. *Interlaced* is useful for the Internet, since it shows parts of the image in the Web browser before the entire image

has been delivered over the (relatively slow) telephone lines. *Non-interlaced* means the entire image must be delivered before it can be displayed.

8. Choose a color for **Transparency**. *Transparency* allows other images or text "underneath" the GIF image to show through the color selected as transparent.

9. Select a **Resolution** for the preview image:

   ➤ **Screen**: Use the screen resolution, typically 72dpi (dots per inch) or 96dpi.

   ➤ **Printer**: Use the printer's resolution, typically 300dpi or 600dpi; use this for the best quality hardcopy.

   ➤ **Source**: Let the destination application figure out the best resolution to use.

   ➤ **Custom**: Specify any resolution; default = 72x72dpi. Visio recommends that the resolution range between 32dpi and 400dpi.

10. Select a **Size** of image.

11. Select a **Color Format**: monochrome (black-white), 16 colors, 256 colors (the default), or 24-bit RGB (16.7 million colors).

12. Select a **Color Translation** format:

    ➤ **Normal**: Keep the colors as they are in the Visio drawing.

    ➤ **Inverse**: Invert the colors (the picture looks like the strip of negatives from color film).

    ➤ **Inverse Grays Only**: Invert only white, black, and grays; keep other colors as they are.

    ➤ **Gray Scale**: Change the colors to levels of gray.

    ➤ **Inverse Gray Scale**: Change the colors to levels of gray and make a negative.

13. When you make changes to the options and wish to save the changes for the next time you use this command, click **New**.

14. Type a name in the **Profile Name** text box. This name will appear in the Profiles list box the next time you see the PNG Output Filter Setup dialog box.

15. Click **OK** to dismiss the dialog box.

16. Click **OK**. Visio converts the drawing and saves it in PNG format.

Export to TIFF

Use the following procedure to convert the Visio drawing to tagged image file format:

1. Ensure a drawing is open in Visio.

2. Select **File | Save As**.

3. Click on the **Save as type** list box and select **TIF** file format.

4. Click **Save**. Notice the TIFF Output Filter Setup dialog box.

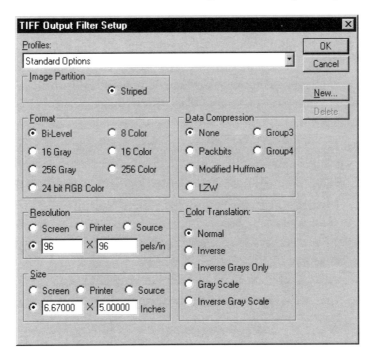

5. Choose a **Profile** from the list box. Visio includes two: Standard Options and Visio HTML Export.

6. Select a **Format**:
   ➤ **Bilevel**: Reduce the image to black and white; creates a smaller file size.
   ➤ **16 Gray**: sixteen shades of gray.
   ➤ **256 Gray**: 256 shades of gray.
   ➤ **8 Color**: eight colors.
   ➤ **16 Color**: sixteen colors.
   ➤ **256 Color**: 256 colors; the default.
   ➤ **24 Bit RGB Color**: 16.7 million colors.

7. Select a **Resolution** for the preview image:
   ➤ **Screen**: Use the screen resolution, typically 72dpi (dots per inch) or 96dpi.
   ➤ **Printer**: Use the printer's resolution, typically 300dpi or 600dpi; use this for the best quality hardcopy.

➤ **Source**: Let the destination application figure out the best resolution to use.

➤ **Custom**: Specify any resolution; default = 72x72dpi. Visio recommends that the resolution range between 32dpi and 400dpi.

8. Choose a **Size**.

9. Select a **Data Compression** for the preview image:

➤ **None**: no compression creates a larger file but may be necessary for some applications that cannot read compressed TIFF files.

➤ **Packbits**: best suited to black-white images.

➤ **Modified Huffman**: best suited for gray images.

➤ **LZW**: best suited for color images.

10. Select a **Color Translation** format:

➤ **Normal**: Keep the colors as they are in the Visio drawing.

➤ **Inverse**: Invert the colors (the picture looks like the strip of negatives from color film).

➤ **Inverse Grays Only**: Invert only white, black, and grays; keep other colors as they are.

➤ **Gray Scale**: Change the colors to levels of gray.

➤ **Inverse Gray Scale**: Change the colors to levels of gray and make a negative.

11. When you make changes to the options and wish to save the changes for the next time you use this command, click **New**.

12. Type a name in the **Profile Name** text box. This name will appear in the Profiles list box the next time you see the TIFF Output Filter Setup dialog box.

13. Click **OK** to dismiss the dialog box.

14. Click **OK**. Visio converts the drawing and saves it in TIFF format.

Export to WMF

Use the following procedure to convert the Visio drawing to Windows metafile format:

1. Ensure a drawing is open in Visio.

2. Select **File | Save As**.

3. Click on the **Save as type** list box and select **WMF** file format.

4. Click **Save**. There are no options for the WMF format. Visio converts the drawing and saves it in WMF format.

Hands-On Activity

In this activity, you use the export functions. Begin by starting Visio.

1. Open document **Perspective Block Diagram.Vsd** supplied in folder **Block Diagram**.

2. Select **File | Save As**.

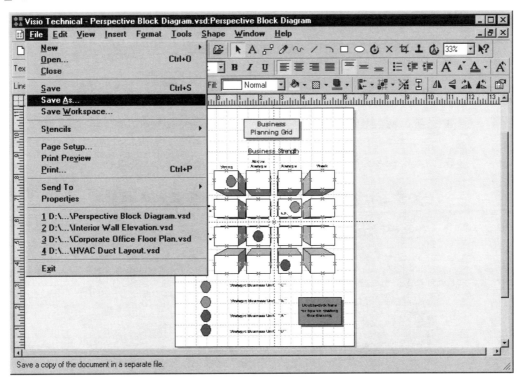

3. Click on the **Save as type** list box and select **GIF** file format.

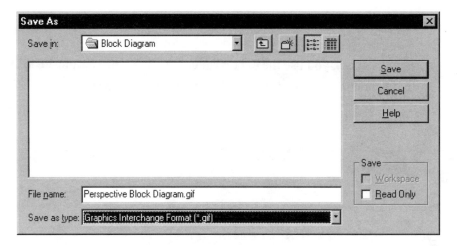

4. If necessary, type a filename in the **File name** text box and select the destination subdirectory.

5. Click **Save**. Notice the GIF Output Filter Setup dialog box.

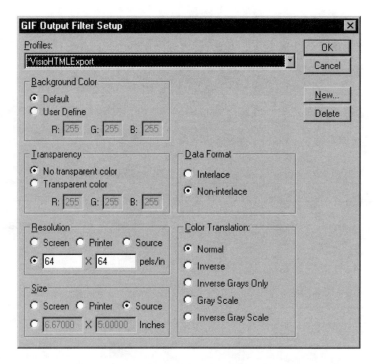

6. Select **Visio HTML Export** option from the **Profiles** list box. This action preselects all other options on your behalf.

7. Click **OK**. Visio converts the drawing and saves it in PCX format.

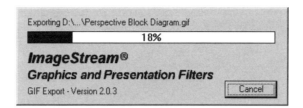

8. Start **Windows Paintbrush** (or another raster editor program) by double-clicking on its icon in the Windows desktop or from the Windows 95 **Start** menu.

9. Select **File | Open** from the menu and open **Perspective Block Diagram.Gif**.

10. Press **Alt+F4** to exit Paintbrush.

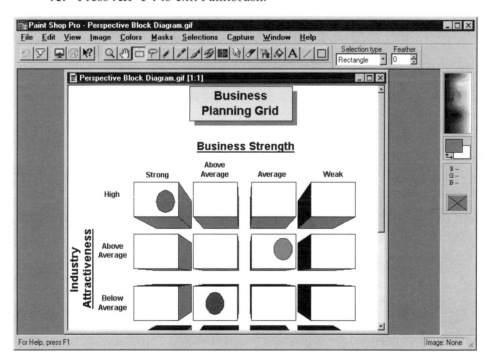

11. Click on Visio and press **Alt+F4** to exit Visio. Click **No** in response to the Save Changes dialog box.

This completes the hands-on activity for exporting Visio drawings in other file formats.

Module 30

Special Selections

Edit | Select Special

Uses

The **Select Special** selection of the **Edit** menu lets you select shapes on the basis of what they are, or by the name of the layer they reside on. In most cases, you click on a shape to select it; to select more than one shape, you hold down the **Shift** key while clicking on shapes. To select all shapes on the page, press **Ctrl+A**.

The **Select Special** command displays a dialog box with several options. You choose shapes and objects in either of two categories: (1) shapes and objects; or (2) layer names.

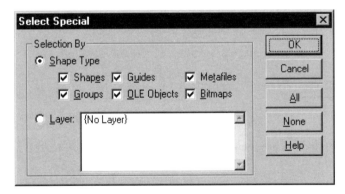

➤ **Shapes**: select all shapes on the current page.
➤ **Groups**: select all grouped objects on the page.
➤ **Guides**: select all guide lines and guide points on the page.
➤ **OLE Objects**: select all linked and embedded objects.
➤ **Metafiles**: select all objects pasted in WMF format.
➤ **Bitmaps**: select all objects pasted in BMP format.

➤ **Layer**: select all objects assigned to a specific layer name.

➤ **All**: select all shapes and objects.

Procedures

Before presenting the general procedures for special selections it is helpful to know about the shortcut keys. These are:

Function	Keys	Menu	Toolbar
Select			↖
Select special		Edit | Select Special	
Select All	Ctrl+A	Edit | Select All	

Using Select Special

Use the following procedure to select objects:

1. Select **Edit | Select Special**. Notice the Select Special dialog box.
2. Choose **Shape Type** or **Layer**.
3. Within the **Shape Type** area, select any combination of shapes, groups, guides, OLE objects, metafiles, or bitmaps.
4. Or, within the **Layer** area, select a layer name from the list box.
5. Click **OK**.

Hands-On Activity

In this activity, you use the special selection function. Begin by starting Visio.

1. Open the document **Basic Network Diagram.Vsd** found in folder **Network Diagrams**.

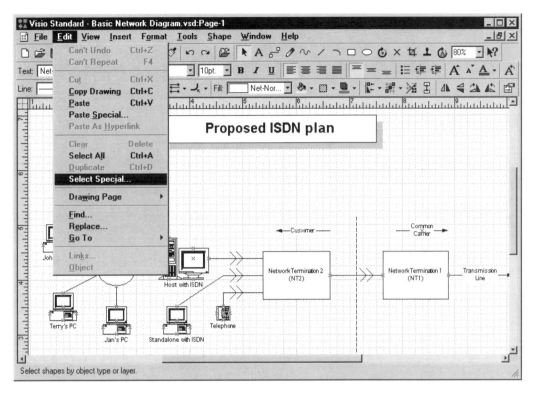

2. Select **Edit | Select Special**. Notice the Select Special dialog box.

3. Click the **None** button to remove all selected options.

4. Click **Shapes**.

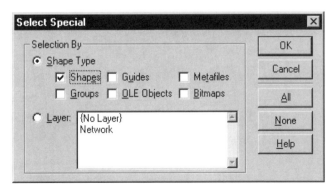

5. Click **OK**. Notice the shapes that are selected, surrounded by cyan handles.

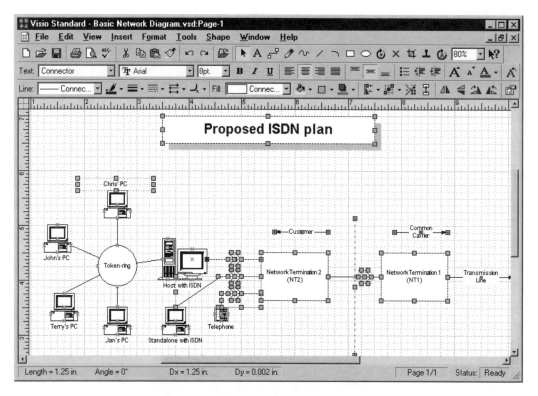

6. Select **Edit | Select Special** again.

7. Click the **None** button to remove all selected options.

8. Click **Groups**.

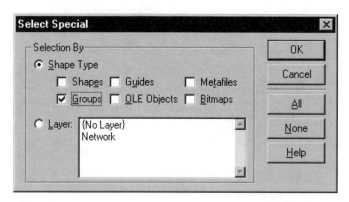

9. Click **OK**. Notice that this time the groups are selected, surrounded by cyan handles.

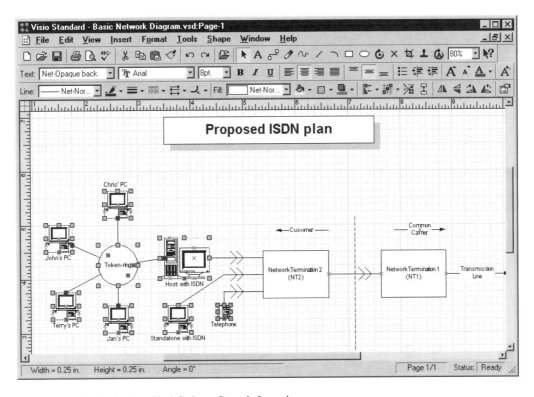

10. Select **Edit | Select Special** again.
11. Click the **Layer** radio button.
12. Click **Network**.

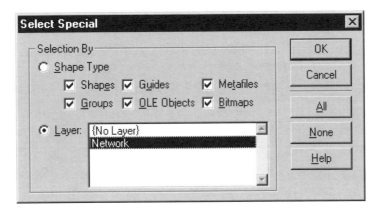

13. Click **OK**. Notice that this time both groups and shapes are selected, but this time only those on layer Network.

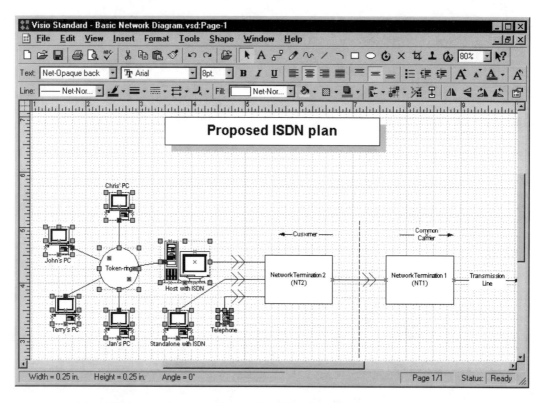

14. Click anywhere on the page to deselect the shapes.
15. Press **Alt+F4** to exit Visio. Click **No** in response to the Save Changes dialog box.

This completes the hands-on activity for special selections.

Module 31

Running Macros
Tools | Macro

Uses

The **Macro** selection of the **Tools** menu is used to run programs external to Visio. While Visio has many drawing and editing functions, it doesn't necessarily do everything you need in your particular discipline. For this reason, Visio lets you write programs that interact with Visio. (Macros were known as "add-on" programs in earlier versions of Visio because they added to the functionality of Visio.)

Some of the macros have been integrated with the Visio menu, such as **File | Open | HTML** and **Shape Explorer**. All of the macro programs included with Visio are listed by the **Tools | Macro** menu selection. The macros provided with your copy of Visio depend on: (1) the version of Visio you have; and (2) any macros your firm may have purchased from third-party developers. All three versions of Visio—Standard, Technical, and Professional—include the programming environment for writing your own macros in VBA (short for Visual Basic for Applications). However, for Visio Standard the documentation for VBA macros is not printed; it is only available in electronic format on the distribution CD-ROM. For more information on writing your own VBA macros, consult *Learn Visio 5.0 for the Advanced User* (Wordware Publishing).

Procedure

Before presenting the general procedures for special selections, it is helpful to know about the shortcut key. This is:

Function	Keys	Menu	Toolbar
Run Macro	Alt + F8	Tools | Macro | Macros	▶

How to Run a Macro

1. Many add-ons require that you select one or more shapes before starting the add-on. Unfortunately, there is no way to tell before you start the macro. Be prepared to cancel the macro, select shapes, and start the macro a second time.

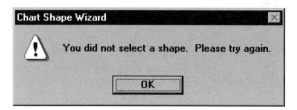

2. There are two ways to start a macro:

   ➤ Select **Tools | Macro**. Notice the submenus that segregate macros by function. The macro names listed vary, depending on the version of Visio you are operating. Visio Standard displays:

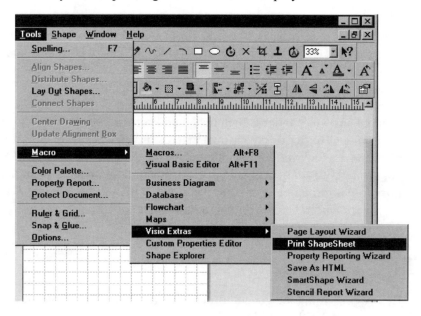

Visio Technical displays:

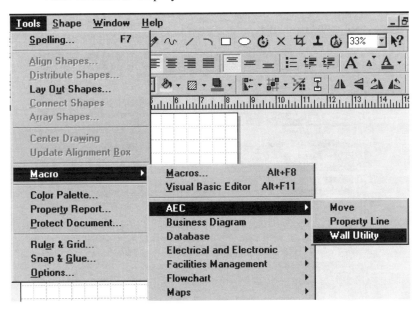

Visio Professional displays:

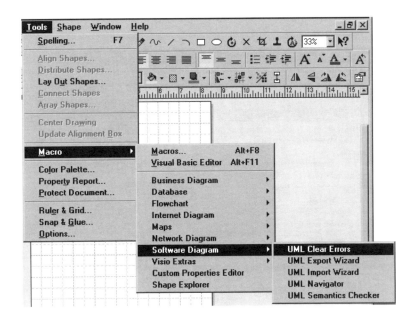

> Select **Tools | Macro | Macro**. Notice the **Macros** dialog box. The list of macros varies, depending on the version of Visio you are operating. Visio Standard displays:

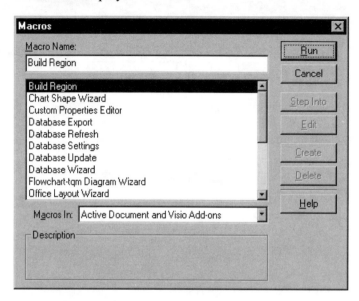

Visio Technical displays:

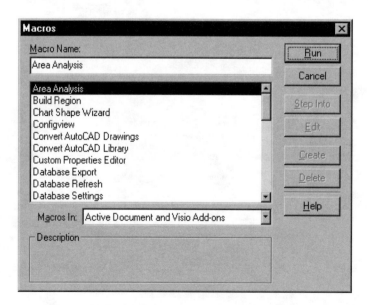

Visio Professional displays:

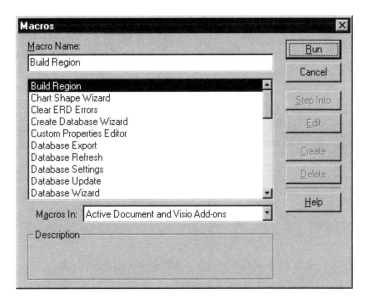

3. Select the name of a macro from the cascading menus or from the list provided in the dialog box.

4. Follow the instructions provided by the add-on's dialog box.

Hands-On Activity

In this activity, you use the Chart Maker add-on. Begin by starting Visio.

1. Open a new, empty drawing.

2. Open the **Basic Shapes.Vss** stencil file from the **Block Diagram** folder.

3. Drag the **Star 5** shape from the stencil into the drawing.

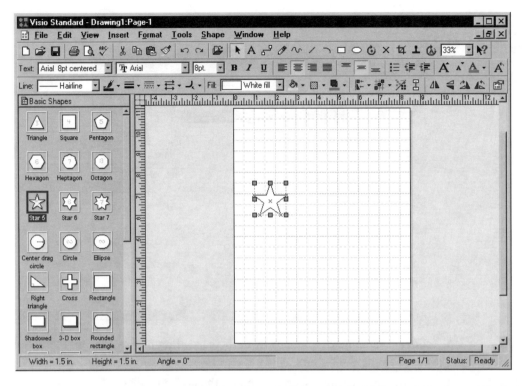

4. Ensure the star shape is selected. Select **Tools | Macro | Business Tools | Chart Shape Wizard**. Notice the start of the Chart Shape Wizard dialog box that lets you create two kinds of graphing shapes: extended shapes and stacked shapes.

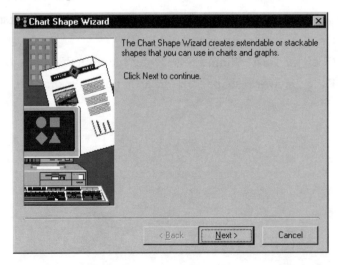

5. Click **Next**.

6. Click **Stackable shape**. Click **Next**.

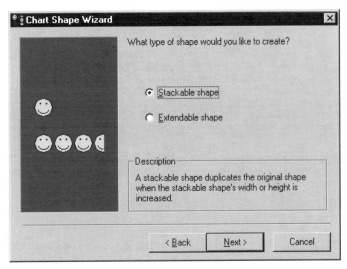

7. Select **Smart Shape**, along with **Horizontal** alignment, **4** shapes, and **0** spacing. Click **Next**.

8. The macro prompts you to select a shape and click **Finish**. Wait while the macro does its work.

9. The chart shape is complete.

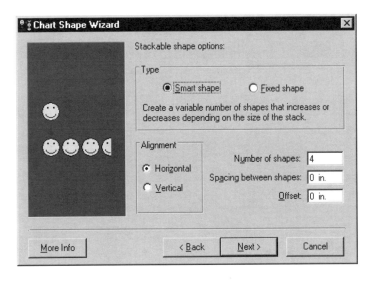

10. Stretch the copied shapes to various lengths.

11. Press **Alt+F4** to exit Visio. Click **No** in response to the Save Changes dialog box.

This completes the hands-on activity for using add-on applications.

Module 32

Double-Clicking the Mouse Button

Format | Double-Click

Uses

The **Double-Click** selection of the **Format** menu specifies the action when you double-click a shape. Most software programs perform a single action (or no action at all) when you double-click an object; Visio lets you choose between as many as ten options from the Double-Click dialog box.

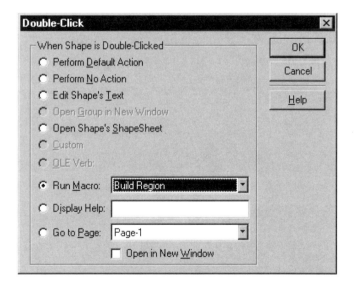

➤ **Perform Default Action**: For a *shape*, allow editing of the shape's text block; for a *group*, open the group in a new window; for an *OLE object*, launch the linked application.

➤ **Perform No Action**: Nothing happens when you double-click.

➤ **Edit Shape's Text**: Applies only to shapes; go into text editing mode.

307

➤ **Open Group in New Window**: Applies only when a group is selected; display the group in an independent window for editing.

➤ **Open Shape's ShapeSheet**: Display the numbers that control the look and size of the shape in a spreadsheet-like interface, as illustrated below.

➤ **Custom**: Perform a custom (user-defined) behavior; available only when a shape's ShapeSheet contains custom properties.

➤ **OLE Verb**: Applies only to an inserted object; typically it is the **Edit** or **Open** command.

➤ **Run Add-on**: Run one of the add-on programs installed with Visio.

➤ **Display Help**: Displays a topic from an HLP help file; you must type **filename.hlp!keyword** or **filename.hlp!#Number** with the following meaning:

Metaname	Meaning
filename.hlp	Name of a Windows help file, such as Shape.Hlp.
!keyword	Index term associated with the help topic, such as "Basic Shape."
!#number	Numeric ID referenced in the map section of the help project file (HPJ).

➤ **Go to Page**: Switch to another page in the drawing.

➤ **Open in New Window**: This option affects all of the above options: open a new window

Procedures

Use the following procedure to change the double-click action assigned to a shape:

1. Select the shape.
2. Select **Format | Double-Click**.
3. Click an option from the Double-Click dialog box.
4. Click **OK**.
5. Double-click on the shape to test the action.

Hands-On Activity

In this activity, you use one of the double-clicking functions. Begin by starting Visio.

1. Open a new, blank drawing.
2. Open the **Symbols.Vss** stencil file from the **Visio Extras** folder.
3. Drag the **Coffee** shape into the drawing.

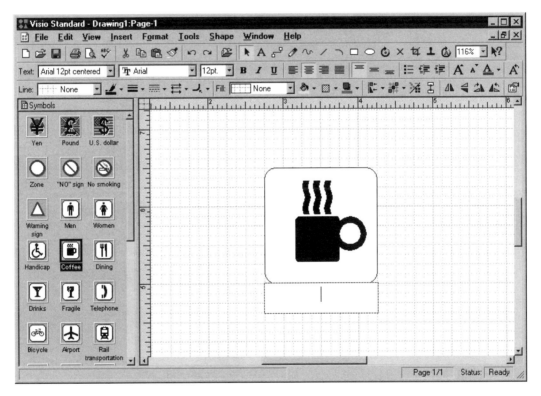

4. Double-click the coffee symbol. Notice that Visio switches to text mode.
5. Select **Format | Double-Click**. Notice that the Double Click dialog box has the Edit Shape's Text option selected.
6. Click **Open Shape's Shapesheet**.

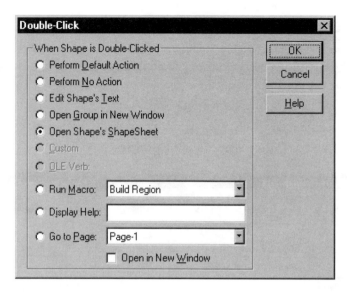

7. Click **OK**.

8. Double-click on the coffee cup. Notice that the shapesheet is displayed in a separate window.

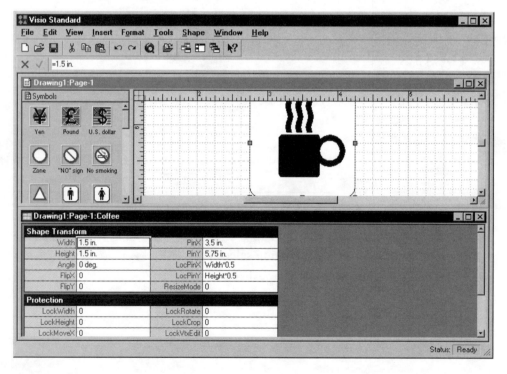

9. Close the shapesheet window.

10. Select **Format | Double-Click**.

11. Click **Open Group in New Window**.

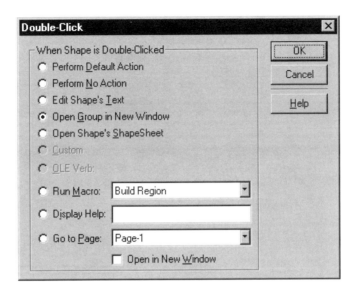

12. Click **OK**.

13. Double-click on the coffee cup. Notice that the group is displayed in a separate window.

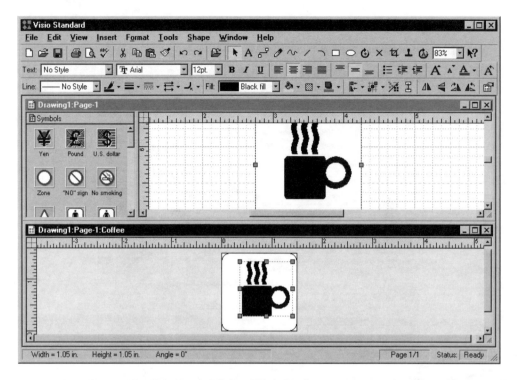

14. Press **Alt+F4** to exit Visio. Click **No** in response to the Save Changes dialog box.

This completes the hands-on activity for changing the double-click action assigned to a shape.

Module 33

Behavior

Format | Behavior, Special

Uses

The **Behavior** and **Special** selections of the **Format** menu are used to change how shapes and groups act and display. The **Behavior** dialog box controls the behavior of each shape. The **Special** dialog box displays basic information and lets you attach data to the shape.

The **Behavior** dialog box controls the interaction, highlighting, resize, and interaction behaviors.

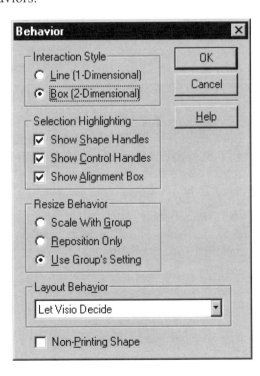

➤ **Interaction Style**: Visio treats lines as one dimensional, while nearly every other shape and object is treated as a two-dimensional object. However, by selecting **Line** for a 2D shape, you change how the shape reacts to resizing and rotation.

➤ **Selection Highlighting**: Determines which of three indicators—shape handles, alignment box, and control handles—display when a object is selected, as the illustration below shows.

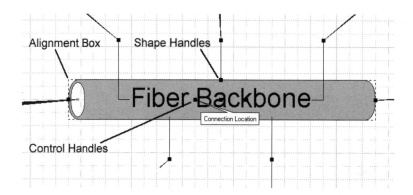

➤ **Resize Behavior**: These three settings affect a shape only when it is part of a group.

➤ **Non-printing Shape**: When this option is turned on, the shape will not appear when the drawing is printed.

➤ **Layout Behavior:** This option specifies how 2D shapes interact with routable connectors. **Let Visio Decide** lets Visio determine when to make the shape *placeable* with dynamic connectors. **Layout and Route Around** forces a 2D shape to always be placeable. **Do Not Layout and Route Around** prevents a shape from being placeable, which prevents it from being detected by routable connectors.

The **Special** dialog box displays some basic information about the shape and lets you attach data to the shape. The data fields, such as **Data 1,** contain a field, which can be inserted with the **Field** command (see Module 24).

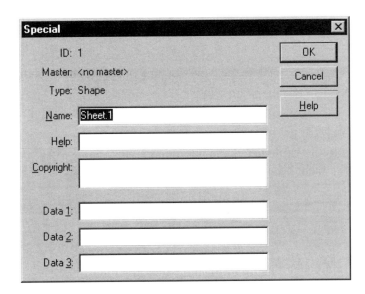

Procedures

Changing a Shape's Behavior

Use the following procedure to change the behavior of a shape:

1. Select a shape.
2. Select **Format | Double-Click** from the menu. Notice the Behavior dialog box.
3. Change one or more options.
4. Click **OK**.
5. Click the shape to test the change in behavior.

Specifying Special Data

Use the following procedure to enter data into a shape:

1. Select a shape.
2. Select **Format | Special** from the menu bar. Notice the Special dialog box.
3. Type new data in the **Name**, **Help**, **Data 1**, **Data 2**, and **Data 3** text boxes, as required.
4. Click **OK**.

Hands-On Activity

In this activity, you use the behavior functions. Begin by starting Visio.

1. Open a new, blank drawing.

2. Open the **Basic Network Shapes.Vss** stencil file, which is found in the **Network Diagram** folder.

3. Drag the **Ethernet** shape into the drawing. Notice its green handles, green alignment box, and blue connection markers.

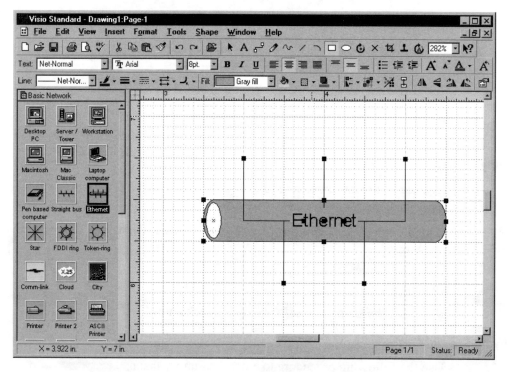

4. Select **Format | Double-Click** from the menu.

5. Turn off all three **Selection Highlighting** options.

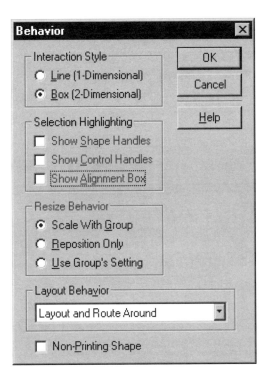

6. Click **OK**. Notice that the selection cues disappear from the shape.

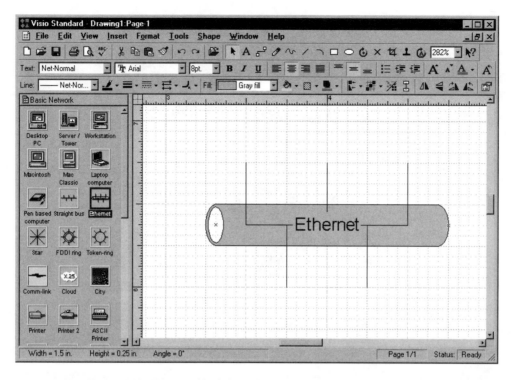

7. Press **Alt+F4** to exit Visio. Click **No** in response to the Save Changes dialog box.

This completes the hands-on activity for changing the behavior of shapes.

Module 34

Custom Properties

Shape | Custom Properties
Tools | Macro | Custom Property Editor, Visio Extras,
Property Reporting Wizard

Uses

The **Custom Properties** selection of the **Shape** menu is used to edit the *custom* (defined by you) properties defined in the shape. The properties are not color or linetype; rather, they are data attached to a shape, such as its name, model number, and price. CAD software sometimes refers to properties as "attributes" or "tag data."

The content of the **Custom Properties** dialog box varies according to the shape and its custom properties. The custom properties are editable. The data itself is edited with the **Custom Properties** command. The data fields are edited with the **Custom Property Editor** command.

You summarize the information using **Property Reporting Wizard**, which places the report on a separate layer in the current drawing page. These reports are excellent for counting all shapes in a drawing, producing a bill of material, or creating an inventory report. Unfortunately, the three Custom Property related tools are not gathered together in one menu, which makes it bit hard to use them.

 Point of Interest: Most shapes do *not* have custom properties defined. In that case, Visio displays a warning dialog box.

319

Procedures

Editing a Shape's Custom Properties

Use the following procedure to edit a shape's custom properties:

1. Select a shape.

2. Select **Shape | Custom Properties**.

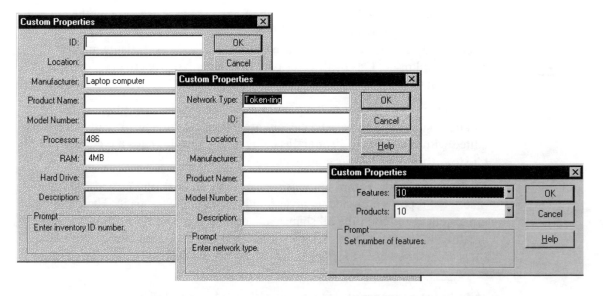

3. Add or change data in the text fields displayed by the Custom Properties dialog box.

4. Click **OK**.

Editing Custom Property Fields

Use the following procedure to edit custom property fields:

1. Select **Tools | Macro | Custom Property Editor** from the menu bar. Notice that Visio loads the Custom Properties Editor.

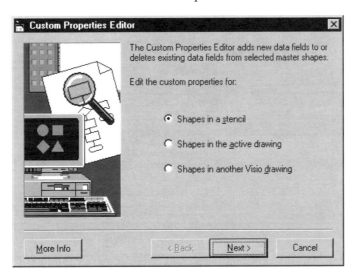

2. The dialog box allows you to edit the custom properties for either: (1) all shapes in a VSS stencil file; (2) all shapes in the current drawing; or (3) shapes stored in another VSD Visio drawing.

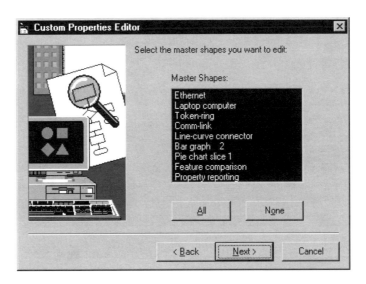

3. Click **Next**. Select the shape(s) you want to edit.

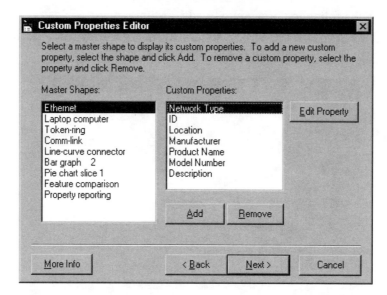

4. Click **Next**. Select the property to edit, add, or remove.

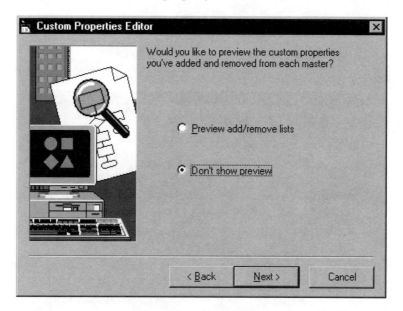

5. Click **Next**. You have a chance to review your changes.

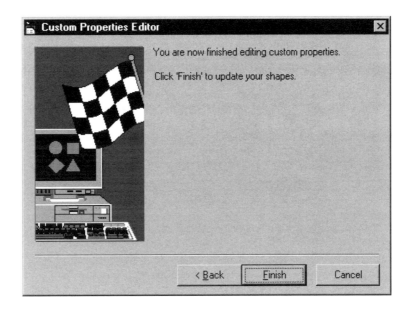

6. Click **Next**. You're almost done.

7. Use the **Custom Properties** command to see the changes to custom properties.

Summarizing Custom Property Data

Use the following procedure to create a report from property fields:

1. Select **Tools | Macro | Visio Extras | Property Reporting Wizard.** Notice the Property Reporting Wizard dialog box.

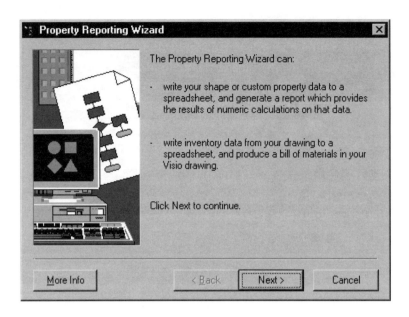

2. Click **Next**. Here you decide which shapes the report is based on:

  ➤ **Range**: In the whole document (when it contains more than one page) or just the current page.

  ➤ **Include**: All shapes, selected shapes, or shapes with specific custom properties.

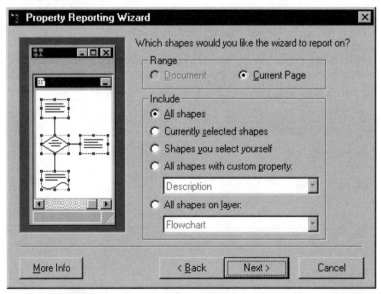

3. Click **Next**. Choose the custom properties you want included in the report.

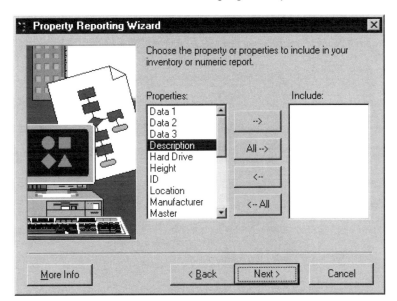

4. Click **Next**. Decide on the kind of calculation you want your report to include. Unfortunately, Visio does do perform all the calculations this dialog box shows.

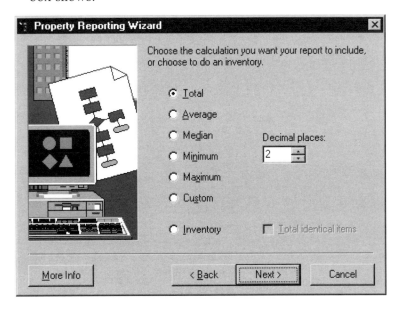

5. Click **Next**. Type in the name for the report and select a drawing page for the report.

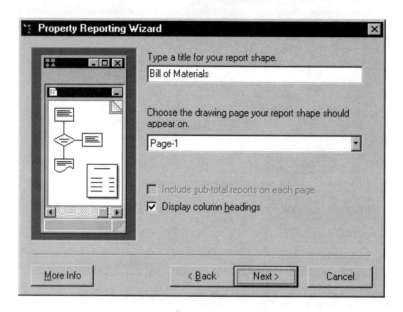

6. Click **Next**. Not done yet. You get to choose if you want even more text in the report. For fun, click all check boxes.

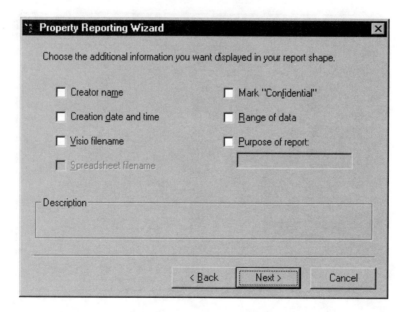

7. Click **Next**. Visio notes you have answered all questions necessary.

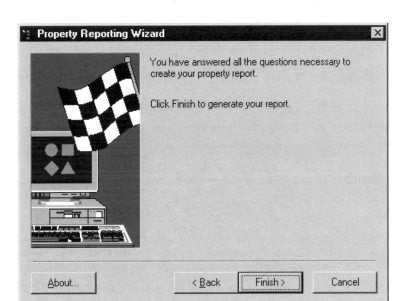

8. Click **Finish**. Watch the blue squares as they march across the dialog box.

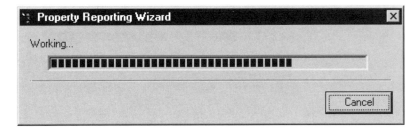

9. From the **Zoom** list box, select **Page** in order to find the property report.

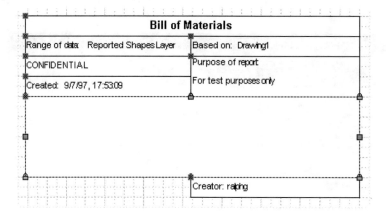

This completes the procedures for working with custom properties.

Module 35

Internet Tools

Insert | Hyperlink
File | Save As | HTML

Uses

Visio allows you to create "Internet" versions of the drawings in two different environments: (1) hyperlinks can be added to the Visio drawing itself, allowing you to jump from document to document on your own computer system; and (2) Visio drawings can be exported in HTML format for viewing by web browsers.

The **Hyperlink** selection of the **Insert** menu places hyperlinks in the Visio drawing. A *hyperlink* is simply a filename with reaction: clicking a hyperlink causes Visio to load the file specified by the hyperlink. A hyperlink is also known as a URL, short for "uniform resource locator," the universal file naming system used by the Internet. Typical hyperlinks (or URLs) look like:

URL (Hyperlink filename)	Meaning
c:\visio\samples\block diagram.vsd	Visio drawing.
c:\visio\solutions\basic blocks.vss	Visio stencil file.
c:\folder\index.htm	HTML document located on your computer
c:\graphics\filename.tif	Graphic file in TIFF format.
http://www.visio.com	Visio's web site.
http://www.wordware.com	Wordware Publishing's web site.
http://users.uniserve.com/~ralphg/	Author Ralph Grabowski's web site.

As you can see from the list, you can hyperlink *any* kind of file in a Visio drawing—whether on your computer or on the Internet. When Visio is unable to display the file, it launches the appropriate application. For example, to display the HTML document, Visio launches your web browser; to display the graphic file, Visio launches your raster editor. Visio searches the Windows registry to determine which program to launch, based on the filename's extension.

Visio drawings and HTML files have one advantage over other kinds of files. In addition to linking to the VSD drawing filename, you can also select the page to display. In the case of HTML files, you can select an internal link name, which are usually prefixed with a # character.

You can hyperlink to Visio drawing, template, and stencil files. In addition, you can create internal hyperlinks, where the Visio drawing links to other parts of itself. When the cursor passes over a hyperlinked shape in a Visio drawing, the cursor changes to show a tiny earth and a three-link chain.

The Visio drawing can be exported to HTML format for viewing by web browsers on your firm's intranet or by everyone on the Internet. The hyperlinks you insert in the drawing are preserved in the HTML document. Each page in the drawing becomes an HTML page. The Visio drawing itself is converted to a raster file, typically in JPEG format. Your Visio drawing cannot be edited when displayed by the web browser.

Procedures

The shortcut keys are:

Function	Keystroke	Menu	Toolbar Icon
Insert Hyperlink	Ctrl+K	Insert | Hyperlink	
Save as HTML	Ctrl+S	File | Save As | HTML	

Most hyperlink-related commands are found on the right-click menu, after you select hyperlinked shape:

Function	Right-click Menu
Display hyperlinked file	Hyperlink | Open
Display in a new window	Hyperlink | Open in New Window
Copy hyperlink to Clipboard	Hyperlink | Copy Hyperlink
Add hyperlink to .URL file	Hyperlink | Add to Favorites
Edit the hyperlink	Hyperlink | Edit Hyperlink

After hyperlinks are added to shapes and pages, these commands navigate hyperlinks within Visio. The keystrokes do not work if the hyperlink points to a file displayed by a program outside of Visio.

Function	*Keystroke*	*Toolbar Icon*
Forward	Alt+Right Arrow	[forward.tif]
Back	Alt+Left Arrow	[back.tif]
Next Page	Ctrl+Page Down	[nextpage.tif]
Previous Page	Ctrl+Page Up	[previouspage.tif]

Add a Hyperlink to a Shape

Use the following procedure to insert a hyperlink in a Visio shape:

1. Select a shape.
2. Select **Insert | Hyperlink** from the menu bar. Notice the Hyperlink dialog box.

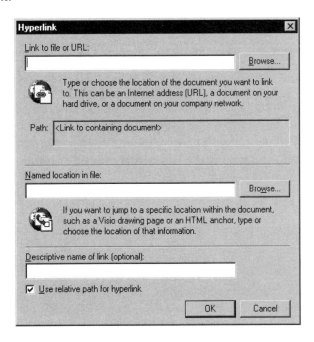

3. There are four areas to fill out in this dialog box, but only the first is required. Type a filename (or URL) in the **Link to file or URL** text box.

4. If you don't remember the filename, click the **Browse** button. Notice the menu listing two choices: **Internet Address** and **Local File**.

Selecting **Internet Address** launches your computer's web browser, such as Netscape Navigator. Selecting **Local File** displays the **Link to File** dialog box, which lets you select a file from your computer or network's disk drives.

5. **Named Location in File**: When the URL points to a Visio drawing or an HTML file, you have the option of specifying a location within the file. In the case of a Visio drawing, you can link to a specific page within the drawing. For HTML files, you can link to a local link.

6. When you cannot remember the page name or link name, click the **Browse** button. Notice that Visio displays the Hyperlink dialog box. Select a page name or HTML link name.

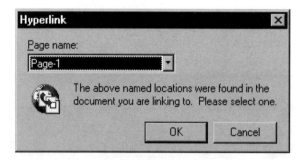

7. When the cursor passes over a hyperlinked shape, the cursor changes to Visio's Internet symbol. In addition, Visio displays a tooltip-like box with the wording you type in the **Descriptive name of link** text box.

8. Turning on **Use relative path** means that you can move the Visio drawing and its linked files together to other folders and drives and the links will still work.

9. Click **OK**. Pass the cursor over the shape to test the link.

Editing a Hyperlink

Use the following procedure to edit the URL and wording of a hyperlink:

1. Move the cursor over shapes to find one with a hyperlink. Notice how the cursor changes shape.

2. Right-click the shape. Notice that the menu includes a Hyperlink item.

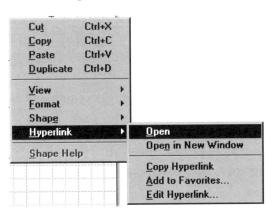

3. Select the **Hyperlink | Edit Hyperlink**. Notice that Visio displays the Hyperlink dialog box.

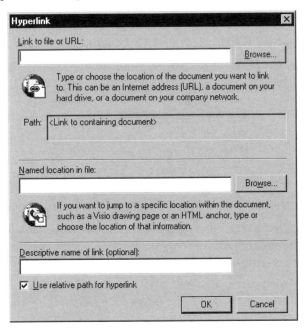

333

4. Change the filename (or URL) in the **Link to file or URL** text box. When you don't remember the filename, click the **Browse** button. Selecting **Internet Address** launches your computer's web browser, such as Netscape Navigator. Selecting **Local File** displays the **Link to File** dialog box, which lets you select a file from your computer or network's disk drives.

5. Change the location within the file with the **Named Location in File** text box. In the case of a Visio drawing, you can link to a specific page within the drawing. For HTML files, you can link to a local link. When you cannot remember the page name or link name, click the **Browse** button. Notice that Visio displays the Hyperlink dialog box. Select a page name or HTML link name.

6. Change the wording in the **Descriptive name of link** text box.

7. Change the setting of **Use relative path**. *Relative path* means that you can move the Visio drawing and its linked files together to other folders and drives and the links will still work.

8. Click **OK**. Pass the cursor over the shape to test the changes you made to the link.

Jumping to a Hyperlink

Use the following procedure to jump to a location specified by a hyperlink:

1. Move the cursor over shapes to find one with a hyperlink. Notice how the cursor changes shape.

2. Right-click the shape. Notice that the menu includes a **Hyperlink** item.

3. Select the **Hyperlink | Open**. Notice that Visio opens the file specified by the hyperlink:

File type	Opened by
Visio file	Visio
HTML file	Web browser
URL	Web browser
Other file	Registered application

When the hyperlink points to another file, you can select either: (1) **Open**, which replaces the current Visio drawing; or (2) **Open in New Window**, which opens the Visio file in another window.

4. Press **Alt+Left Arrow** to move back to the original file. The keystroke does not work if the hyperlink points to a file displayed by a program outside of Visio.

Exporting to an HTML File

Use the following procedure to export the Visio drawing in HTML format:

1. Select **File | Save As** from the menu bar. Notice the Save As dialog box.
2. From the **Save as type** list box, select **HTML Files (*.htm,*.html)**.
3. If necessary, specify a different filename, select another folder and drive.
4. Click **Save.** Notice the Save as HTML dialog box.

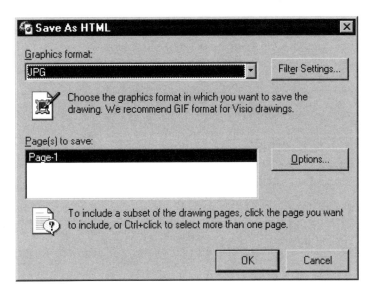

5. Select a **Graphics Format**. The Visio drawing is *not* retained in Visio format when it is exported to HTML; rather, it is converted to a raster (bitmap) image.

   Although you have a choice of six raster formats (GIF, JPG, PNG, TIF, PCX, and BMP), you should always choose JPG for two reasons: (1) it does the best job of compression; and (2) it is understood by all web browsers. There is one exception to choosing JPG: it does lossey compression, which means that fine lines found in Visio drawings might become blurry. In that case, choose PNG format.

6. Click **Filter Settings**. Notice that Visio displays the Filter Settings dialog box.

335

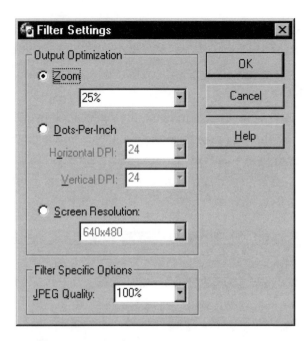

7. The **Output Optimization** section lets you select how the Visio drawing will appear in the HTML document. You have three options:

➤ **Zoom**: reduce (such as 25%; the default) or enlarge (such as 200%) the drawing from its current size.

➤ **Dots-Per-Inch**: specify the resolution in horizontal and vertical dpi, ranging from 24dpi to 96 dpi.

➤ **Screen Resolution**: select a target screen resolution, ranging from 640x480 to 1600x1280.

8. The bottom of the dialog box contains an option specific to the file format:

Format	Filter Specific Option
BMP	None
GIF	Interlacing
JPEG	Quality (compression)
PCX	None
PNG	None
TIFF	None

9. Click **OK**.

10. **Page(s) to Save:** When the Visio drawing contains more than one page (and most don't), you can select which pages you want exported. These pages will be automatically hyperlinked in the HTML document.

11. Click **Options**. Notice that Visio displays the Export Options dialog box.

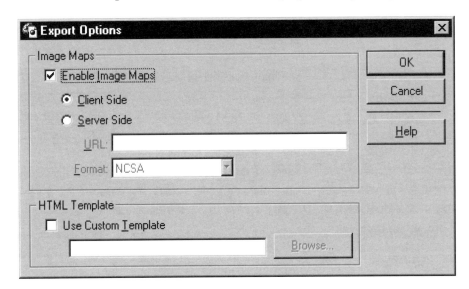

12. *Image maps* are those images you see at a web site that has one or more areas you click to hyperlink somewhere. When your Visio drawing contains hyperlinks, it automatically becomes an image map. Unless you have a specific reason for disabling the hyperlinks you inserted in the Visio drawing, keep **Enable Image Maps** turned on (the default).

13. Image maps work two ways: client or server. The *client* is your web browser; the *server* is the web site that serves up your HTML sites. In most cases, you want the web browser to handle the hyperlinking of image maps. However, in some corporate environments, the web server is in charge of image maps. In that case, select **Server Side** and ask your Network Manager for the **URL** and **Format** data.

14. By default, Visio exports the drawing to a generic-style HTML page. However, you can instruct Visio where to place elements in the HTML page via a template file. The template is an HTML file that contains *substitution codes*. Click **Use Custom Template** and specify the filename. Click **Browse** to search for the file.

Visio recognizes the following substitution codes to customize the output from the **Save As HTML** command. Note that <! ... > is the HTML tag for a comment. Thus, substitution codes are ignored by web browsers. Note that Visio retrieves some of the information from the **File | Properties** dialog box. Create this file with a text editor, giving it an HTM extension.

Codes	Meaning
<!—IMAGE— >	The Visio drawing.
<!—CS_IMAGE_MAP— >	Map data containing hyperlinks.
Navigation Buttons	
<! —NEXT_ANCHOR— >	Makes the **Next** button a navigational link to the next page.
<!—NEXT_PAGE_BTN— >	Tag for a **Next** button graphic. (<IMG SRC="Next.GIF" >)
<!—NEXT_ANCHOR_END >	Ending anchor tag for the **Next** button.
<! —BACK_ANCHOR— >	Make the **Back** button a navigational link to the previous page.
<!—BACK_PAGE_BTN— >	**Back** button graphic. (<IMG SRC="Back.GIF" >)
<!—BACK_ANCHOR_END >	Ending anchor tag for the **Back** button.
**File	Properties Data**	
<!—FILE_NAME— >	Visio drawing filename.
<!—FILE_PATH— >	Visio drawing pathname.
<!—FULL_NAME— >	Visio drawing drive, path, and filename.
<!—CREATOR— >	Creator.
<!—DESCRIPTION— >	Description.
<!—KEYWORDS— >	Keywords.
<!—SUBJECT— >	Subject.
<!—TITLE— >	Title.
<!—*x*— >	*x* = Category, Company, Manager, or Hyperlink_Base.
Page Numbering	
<!—PAGE_COUNT— >	Total number of pages in the Visio drawing.
<!—PAGE_INDEX— >	Page number relative to other pages in the Visio drawing.
<!—PAGE_NAME— >	Name of Visio page saved in HTML format.
<!—HTML_PAGE_COUNT— >	Total number of HTML pages.
<!—HTML_PAGE_INDEX— >	HTML page number relative to other HTML pages.

15. Click **OK**.
16. Click **OK**. Notice that Visio offers to let you view the HTML file.

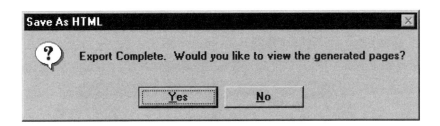

17. Click **Yes**. Notice that Visio launches your web browser and loads the just-converted HTML file.

Hands-On Activity

In this activity, you insert a hyperlink and export the drawing in HTML format. Start by ensuring that Visio is running.

1. Select **File | New | Browse Sample Drawings**.

2. Open the **Global Organization Chart.Vsd** drawing file found in the **Business Diagram** folder.

3. Select **Edit | Go to | 3 VP Sales**.

4. Select **View | Page Width.**

5. Click the **VP Sales** shape. Notice that Visio highlights it.

6. Select **Insert | Hyperlink** from the menu bar. Notice that Visio displays the Hyperlink dialog box.

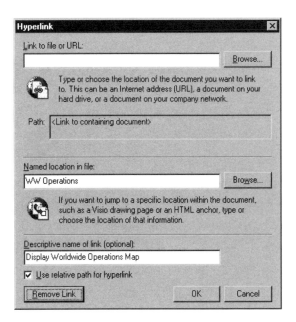

7. We will link back to the first page in this drawing. Press the **Tab** key to move to the **Browse** button next to **Named location in file.** Notice that Visio displays the following:

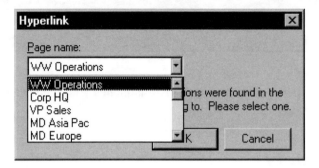

8. Click the **Page name** list box and select **WW Operations**.

9. Click **OK**. Notice that Visio fills in the name for you.

10. Type **Display Worldwide Operations Map** in the **Descriptive Name** text box.

11. Click **OK**.

12. Pass the cursor over the shape. Notice how the cursor changes shape and the "Display Worldwide Operations Map" tooltip.

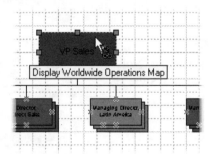

13. Right-click the shape and select **Hyperlink | Open**. Notice that Visio displays page WW Operations.

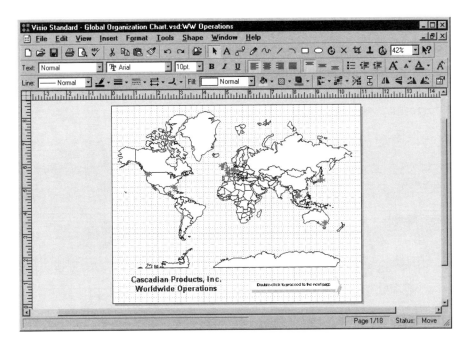

14. Select **File | Save As** from the menu bar. Notice the **Save As** dialog box.

15. From the **Save as type** list box, select **HTML Files (*.htm,*.html)**.

16. If necessary, specify a different filename, select another folder and drive.

17. Click **Save.** Notice the Save as HTML dialog box.

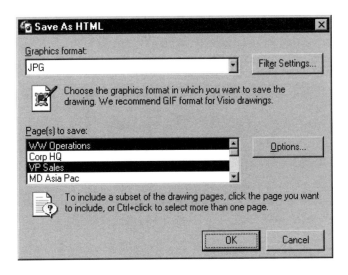

18. Select **JPG** for the **Graphics format**.

19. Select **WW Operations** from **Page(s) to save**.

20. Hold the **Ctrl** key and select **VP Sales**.

21. Click **OK**. Notice that Visio translates the two Visio drawing pages to JPG format.

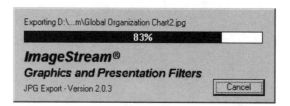

22. If Visio complains that the action cannot be completed, click the **Retry** button and wait a bit longer.

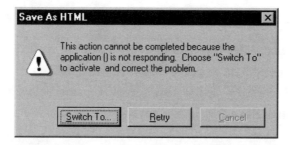

23. When the export is complete, click **Yes** to view the HTML pages. Notice that Visio launches your computer's web browser.

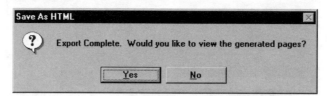

24. In the web browser, press the large gray > button to see the second page of the Visio drawing.

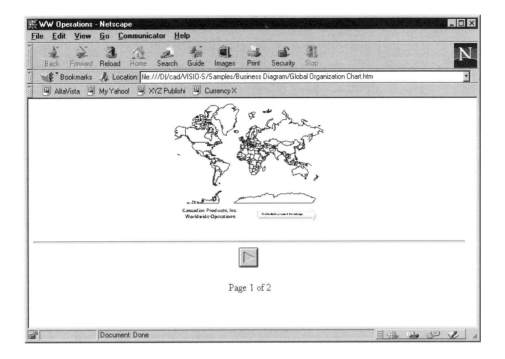

25. Pass the cursor over the green "VP Sales" box. Notice that the web browser displays a tooltip reading "VP Sales."

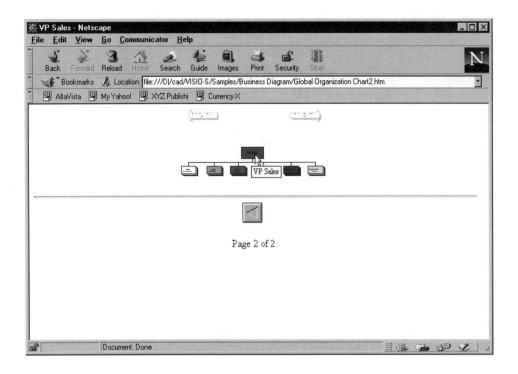

26. Click the **VP Sales** box. Notice that the web browser displays the first page again.

27. Exit Visio and the web browser with **Alt+F4**.

This completes the hands-on activity for creating and using hyperlinks.

Module 36

Mapping a Web Site
Web Diagram Wizard (Visio Professional only)

Uses

Visio allows you to create drawings that map a web site in two ways: (1) manually, using the **Web Diagram Shapes** stencil; and (2) automatically, via the **Web Diagram Wizard**. These two features are only found in Visio Professional. Since the URLs are saved as hyperlinks in the Visio drawing, the drawing can be converted to an HTML file and used as a map at your web site. See Module 35 for information on how to do this.

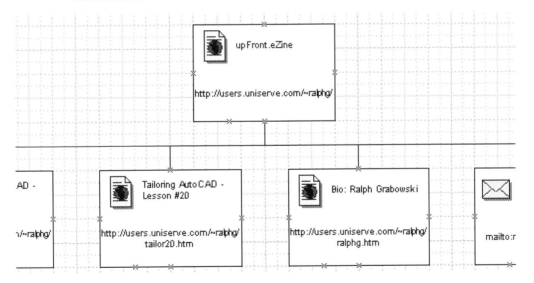

The **Web Diagram Shapes** essentially consist of rectangular shapes (that represent URLs or web pages) and connectors that show the links between the web pages. To modify the data stored by the shapes, right-click to display a context-sensitive menu.

The **Web Diagram Wizard** reads all of the hyperlinks it finds in HTML documents stored *locally* on your computer or *remotely* at a web site. The wizard then creates a drawing showing the HTML pages and the links between them. Unfortunately, you cannot use the Visio drawing to update your web site.

Procedures

The shortcut key is:

Function	Menu
Map Web Site	Tools | Macro | Internet Diagram | Web Diagram Wizard

Create a Web Site Drawing

Use the following procedure to draw a web site:

1. Open the **Web Diagram Wizard** stencil file, found in the **Internet Diagram** folder.

2. Drag shape **URL** into the drawing. Notice the text areas, which must be edited.

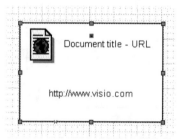

3. Right-click the shape. Notice the menu.

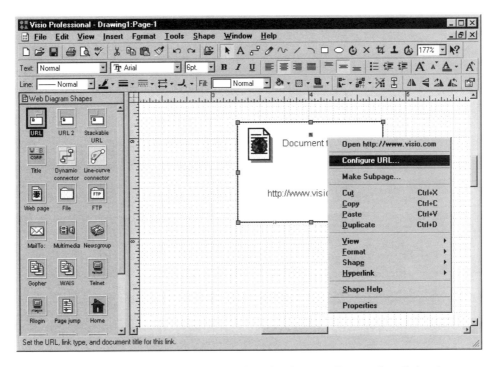

4. Select **Configure URL**. Notice the Custom Properties dialog box.

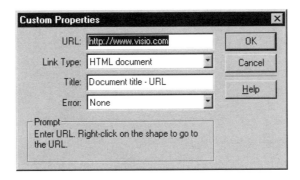

5. Type the URL in the **URL** text box. URL is short for "uniform resource locator" and is the universal file naming system for the Internet. This is displayed in the upper half of the URL shape.

6. Select a **Link Type** from the list box. You can chose from:

347

Link Type	Meaning
File	A file on your computer or local network
FTP	File Transfer Protocol
Gopher	A structured Internet site
HTML document	Hypertext Markup Language
Mailto:	E-mail address
Multimedia	An audio or movie file
Newsgroup	The name of a USENET (news group)
Other	Miscellaneous link
Telnet	A text-based interface to the Internet
WAIS	Wide Area Information Service

7. Type a name for the URL **Title**. This is displayed in the lower part of the URL shape.

8. Select an **Error** from the list box. Normally, you select "None" unless you know that an error exists: Failed to load, Site not found, or Site timed out.

9. Click **OK**. Notice that Visio updates the text of the URL shape.

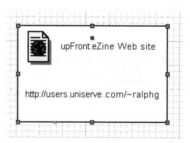

10. Drag additional URL shapes into the drawing.

11. Drag connectors to show links between the URLs.

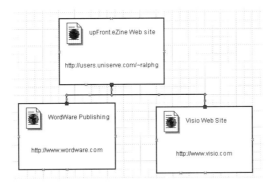

12. Save your work.

Mapping a Web Site

Use the following procedure to automatically map a web site. If the web site is not located on your computer, you must be connected to a network or the Internet for this procedure to work.

1. Start Visio with a new, blank drawing.

2. Select **Tools | Macro | Internet Diagram | Web Diagram Wizard.** Notice that Visio starts the wizard.

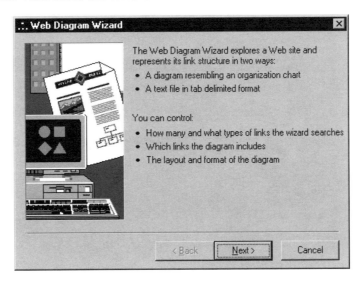

3. Click **Next**. The first time you use the wizard, you must read data from a web site. During that process, Visio creates a text file with formatted data that represents the web site. The second time you use the wizard, you have the option of reading that text file, which is much faster than going through the mapping process a second time.

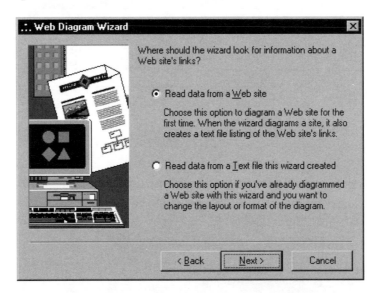

4. Click **Next**. Type the URL for the web site. If the web site is located on your computer, you can type the file specification.

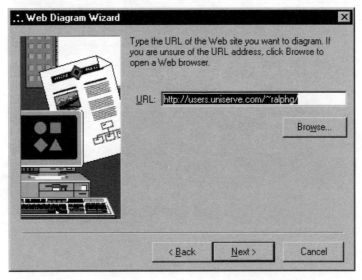

5. Click **Next**. Notice that Visio now launches your web browser with the URL.

6. Click **OK** when the browser has connected with the URL. Notice that Visio takes a minute or two to read the home page.

7. Decide if you want Visio to limit the depth and width of links it finds.

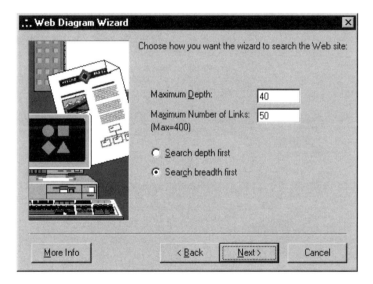

8. Click **Next**. Decide whether you want Visio to check local (internal to the web site) and/or remote web sites.

351

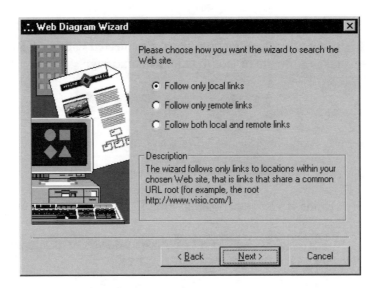

9. Click **Next**. Select the types of links you want Visio to map.

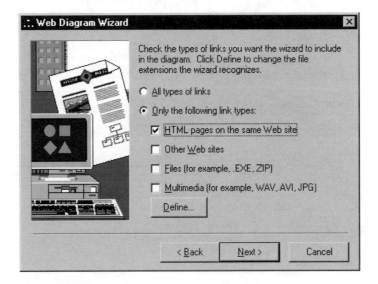

10. Click **Next**. Decide the orientation of the URL shapes.

11. Click **Advanced**. Specify the spacing between URL shapes.

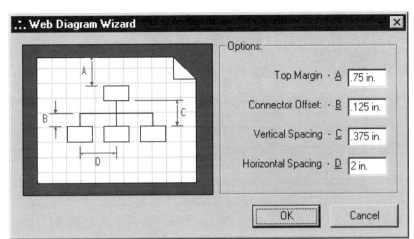

12. Click **OK**. Click **Next**. Decide whether the entire web map should fit on a single page or break up the map to several pages.

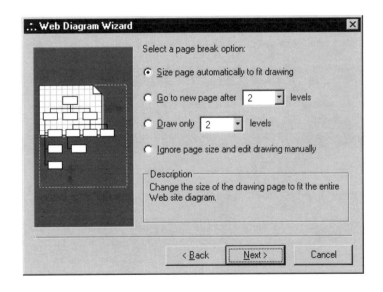

13. Click **Next**. Select the shape of URL shapes and connectors.

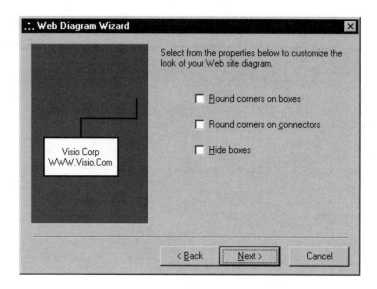

14. Click **Next**. You have completed answering the wizard's many questions.

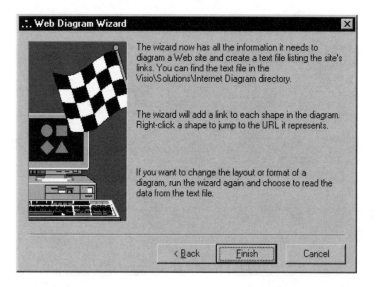

15. Click **Finish**. Sit back and wait while Visio follows the many dozens of links that make up a typical web site. For a modest web site, such as my own, the process takes about 15 minutes.

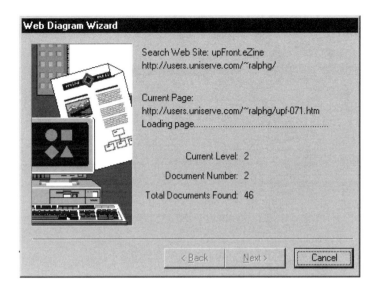

Once Visio has completed collecting data, it begins constructing the drawing.

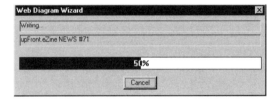

The drawing of a web site map can be quite large, as shown in the illustration on the next page.

355

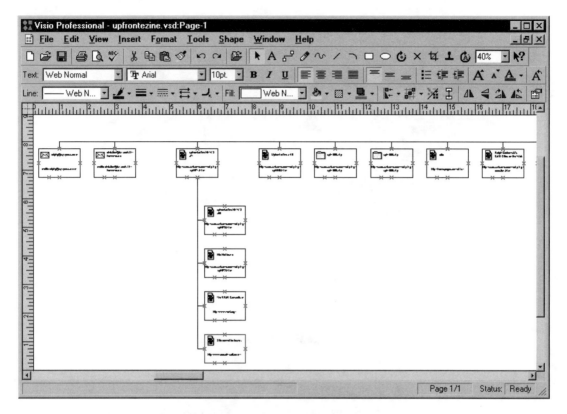

16. Save your work.

This completes the procedures for creating maps of a web site.

Module 37

AutoCAD Drawing Import-Export

(Visio Technical Only)

Uses

AutoCAD is the most popular CAD (computer aided design) program in use to-day. It saves its drawings in a file format known as DWG (short for "drawing"). Visio Technical is able to read and write DWG files created by AutoCAD Release 13 and earlier. In addition, it can read and write files in DXF (short for "drawing interchange format"), a file format developed for AutoCAD but popularly used to exchange drawings between AutoCAD and other CAD programs.

CAD drawings contain a great deal of information, in addition to the lines and circles that make up a drawing. The color, line style, and layer can contain legal information. Unfortunately, the translation of a CAD drawing to a foreign program usually results in the loss of some data and the modification of other data. AutoCAD objects are erased when Visio is unable to convert them.

For this reason, Visio Technical provides two methods of importing an AutoCAD drawing: (1) display only; and (2) editable. The display-only import is more likely to be visually accurate but cannot be edited by Visio. The editable import can be edited by Visio but may have objects missing, is limited to Release 12 (or earlier), and is extremely slow. A moderate size AutoCAD drawing can take a half-hour to import, compared with 30 seconds to load the same drawing into AutoCAD.

Whether imported as display-only or editable, Visio erases the following objects from an AutoCAD drawing:

Display-only Import (R13 and earlier)	Editable Import (R12 and earlier)
2D region	Attributes
3D ACIS solids	3D object of any kind
Externally referenced drawing	Externally referenced drawing

Display-only Import (R13 and earlier)	Editable Import (R12 and earlier)
Leader	Point
Proxy (zombie) object	Variable-width polyline
Shape	Shape
Shapes in complex linetype	
Tolerance dimension	

All layer data is preserved. All fonts are translated to the Arial TrueType font. Visio does not read AutoCAD's SHX fonts nor text styles, such as obliquing and width factor. Even though text style and linestyle names are imported, they lose their properties. For example, the dashed linetype becomes a continuous line.

Procedures

As you can see from the above table, translation is *never* 100 percent accurate. With this caveat in mind, here are the procedures for importing and exporting AutoCAD DWG files.

Import Dwg Display-Only

Use the following procedure to import an AutoCAD drawing into Visio for viewing only:

1. Select **File | Open** from the menu bar. Notice the Open dialog box.

2. From the **Files of type** list box, select **AutoCAD (*.dwg,*.dxf)**. Notice that the importing of DXF files is identical to DWG.

3. Select the AutoCAD drawing file. If necessary, change folders and drives.

4. Click **Open**. Notice the AutoCAD Drawing Layer Status dialog box. It is here that you decide whether to import the drawing as display-only or as editable—or as a combination of the two.

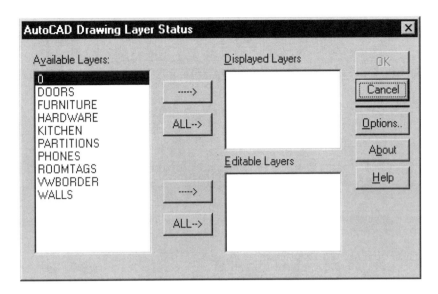

5. To import the drawing as display only, click the **All** button next to **Displayed Layers.**

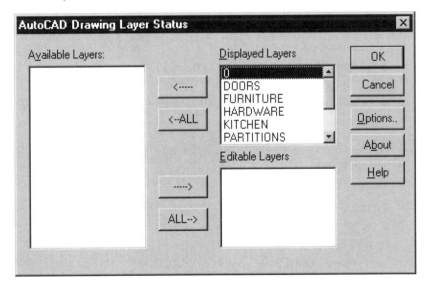

6. Click **Options**. Notice the AutoCAD Converter Options dialog box.

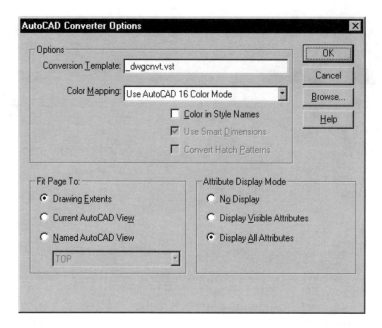

7. Select from the **Options** available:

   ➤ **Conversion Template:** Name of the Visio drawing template VST file that contains preset preferences, such as page size, orientation, etc.

   ➤ **Color Mapping**: choose from monochrome (black-white), 16-color (the default), or 256-color mode (the recommended setting).

   ➤ **Color in Style Names**: retain AutoCAD color numbers when styles are imported.

The other two options are grayed out because they are not available for display-only importation.

8. Select from the **Fit Page To** options:

   ➤ **Drawing Extents** is the default and the best choice for most drawings.

   ➤ **Current AutoCAD View** is the view when the drawing was last saved in AutoCAD.

   ➤ **Named AutoCAD View** lets you select from named views that might have been created in AutoCAD. This option is important when importing 3D drawings.

9. Decide whether you want AutoCAD attributes displayed:

   ➤ **No Display** suppresses the importation of attributes. This can be a good thing since visible attributes can clutter a drawing; in any case, attributes cannot be used when they are displayed in Visio.

➤ **Display Visible Attributes** does just what it says; invisible attributes are not imported.

➤ **Display All Attributes** (the default) displays attributes that are both visible and invisible.

10. Click **OK** to return to the AutoCAD Drawing Layer Status dialog box.

11. Click **OK**. Notice the Page Size/Scale dialog box.

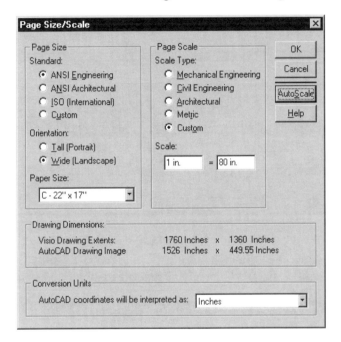

12. By default, Visio imports AutoCAD drawings onto a C-size page in landscape orientation. With this dialog box, you can change the page size, the page orientation, and specify the scale. The fastest method is to click the **AutoScale** button.

13. Select the units that the AutoCAD drawing units should be interpreted as. The default is inches.

14. Click **OK**. Notice that Visio then begins translating the AutoCAD drawing.

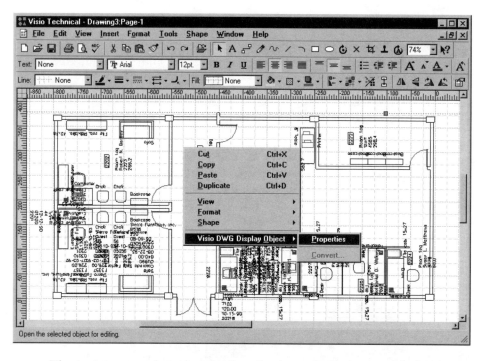

The text you see cluttering the drawing is attribute data made visible.

15. Right-click the drawing and select **Visio DWG Display Object | Properties**. Notice the Visio DWG Display Properties tabbed dialog box.

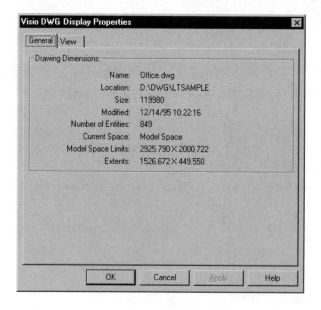

16. The **General** tab provides some information about the AutoCAD drawing.

17. Click the **View** tab. This lets you toggle layers on and off, select a named view, change the units, and scale. When the drawing contains 3D views, Visio will not display the drawing since it does not import 3D data.

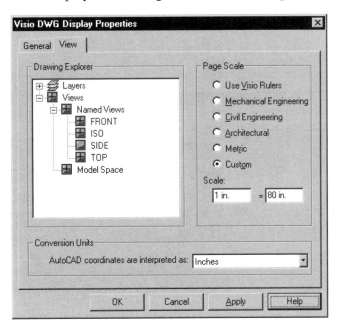

18. To turn a layer off, double-click its name. Then click the **Apply** button to see the effect.

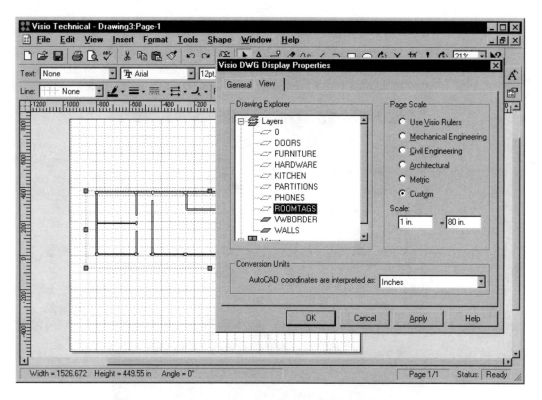

19. While you can draw over the AutoCAD drawing, you cannot edit the drawing itself.

Import Dwg Editable

Use the following procedure to import an AutoCAD drawing into Visio for editing:

1. Select **File | Open** from the menu bar. Notice the Open dialog box.

2. From the **Files of type** list box, select **AutoCAD (*.dwg,*.dxf)**. Notice that the importing of DXF files is identical to DWG.

3. Select the AutoCAD drawing file. If necessary, change folders and drives.

4. Click **Open**. Notice the AutoCAD Drawing Layer Status dialog box. It is here that you decide whether to import the drawing as display-only or as editable—or as a combination of the two.

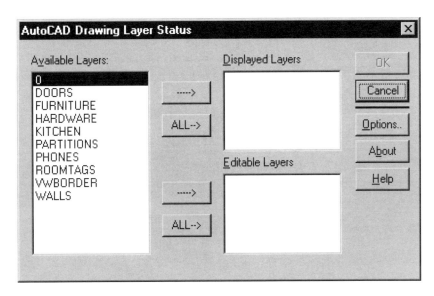

5. To import the drawing as editable, click the **All** button next to **Editable Layers.**

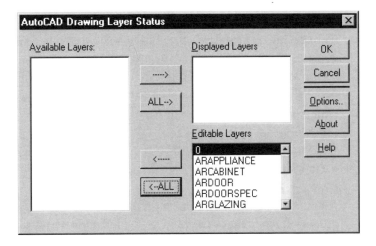

6. Click **Options**. Notice the AutoCAD Converter Options dialog box.

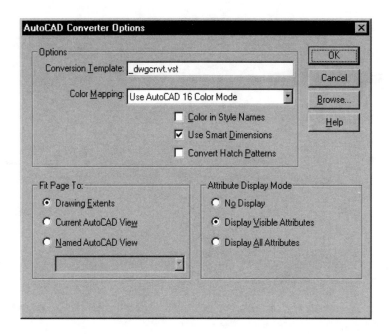

7. Select from the **Options** available:
   - ➤ **Conversion Template:** Name of the Visio drawing template VST file that contains preset preferences, such as page size, orientation, etc.
   - ➤ **Color Mapping**: choose from monochrome (black-white), 16-color (the default), or 256-color mode (the recommended setting).
   - ➤ **Color in Style Names**: retain AutoCAD color numbers when styles are imported.
   - ➤ **Use Smart Dimensions**: converts AutoCAD's associative dimensions to Visio's smart dimensions.
   - ➤ **Convert Hatch Patterns**: since Visio converts each line and dot of an AutoCAD hatch pattern into individual objects, it is best to leave this option off since that saves a great deal of time and disk space.
8. Select from the **Fit Page To** options:
   - ➤ **Drawing Extents** is the default and the best choice for most drawings.
   - ➤ **Current AutoCAD View** is the view when the drawing was last saved in AutoCAD.
   - ➤ **Named AutoCAD View** lets you select from named views that might have been created in AutoCAD. This option is important when importing 3D drawings.

9. Decide whether you want AutoCAD attributes displayed:

   ➤ **No Display** suppresses the importation of attributes. This can be a good thing since visible attributes can clutter a drawing; in any case, attributes cannot be used when they are displayed in Visio.

   ➤ **Display Visible Attributes** does just what it says; invisible attributes are not imported.

   ➤ **Display All Attributes** (the default) displays attributes that are both visible and invisible.

10. Click **OK** to return to the AutoCAD Drawing Layer Status dialog box.

11. Click **OK**. Notice the Page Size/Scale dialog box.

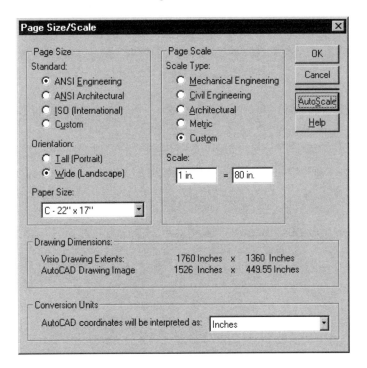

12. By default, Visio imports AutoCAD drawings onto a C-size page in landscape orientation. With this dialog box, you can change the page size, the page orientation, and specify the scale. The fastest method is to click the **AutoScale** button.

13. Select the units that the AutoCAD drawing units should be interpreted as. The default is inches.

367

14. Click **OK**. Notice that Visio then begins translating the AutoCAD drawing. The process can take a long time and depends on the number of objects in the AutoCAD drawing and the speed of your computer. This is a good time to get caught up in answering your e-mail.

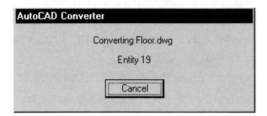

Notice that Visio extracts AutoCAD's blocks (symbols) and creates a stencil file containing the blocks as shapes.

Export to Dwg or Dxf

Use the following procedure to convert the Visio drawing to AutoCAD drawing format *(available in Visio Technical only)*:

1. Ensure a drawing is open in Visio.

2. Select **File | Save As**.

3. Click on the **Save as type** list box and select **DWG/DXF** file format.

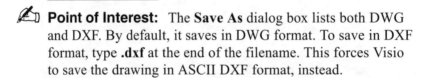 **Point of Interest:** The **Save As** dialog box lists both DWG and DXF. By default, it saves in DWG format. To save in DXF format, type **.dxf** at the end of the filename. This forces Visio to save the drawing in ASCII DXF format, instead.

4. Click **Save**.

5. Notice the Visio Drawing Layer Status dialog box.

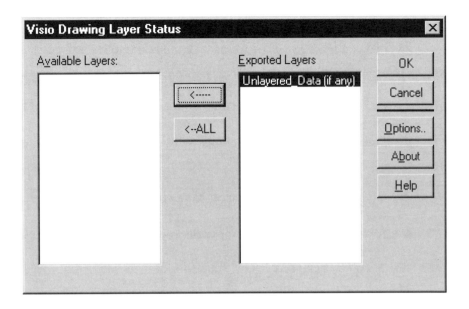

6. Click **ALL** to export shapes on all layers. In many cases, a Visio drawing has only one layer. When it has more than one layer, you have the option of selecting the layers you want exported.

7. Click **Options**. Notice the Visio Exporter Options dialog box.

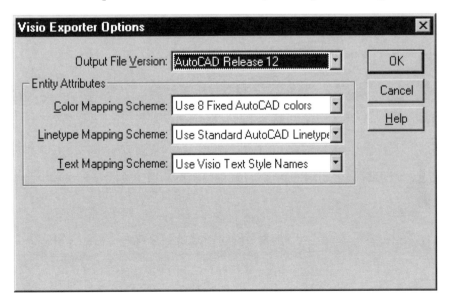

8. Select an **Output File Version**. The **AutoCAD Release 12** output file version works with most copies of AutoCAD, including Release 11, 12, and 13, all versions of AutoCAD LT, and most CAD packages that claim to read DWG files.

9. Select options for **Entity Attributes**. In all three cases, the default works satisfactorily:

   ➤ **Color Mapping Scheme**: choose between no color (black-white), 8 fixed colors (the default), 16-color mode, or 256-color mode.

   ➤ **Linetype Mapping Scheme**: choose between no linetypes (all lines are continuous), translate to AutoCAD's standard linetypes (the default), or keep Visio line style names.

   ➤ **Text Mapping Scheme**: choose between keeping Visio text style names (the default) or removing all text style names.

10. Click **OK** to dismiss the dialog box.

11. Click **OK**. Visio converts the drawing and saves it in DWG format.

This concludes the procedures for converting AutoCAD drawings to and from Visio.

Appendix

Keystroke Shortcuts and Toolbar Icons

Keystroke Shortcuts

This appendix lists all of Visio's keystroke shortcuts in alphabetical order twice. First in alphabetical order of the tool name; then in alphabetical order by keystroke.

Function	Keystroke
Activate the menu bar	Alt or F10.
Actual Size	Ctrl+1
Add to selection set	Shift
Align Shapes	F8
Arc Tool	Ctrl+7
Bring Shape To Front.	Ctrl+F
Cascade Windows	Alt+F7
Connect Shapes	Ctrl+K
Connector Tool	Ctrl+3
Constrain Movement	Shift
Copy to Clipboard	Ctrl+C
Copy Shape	Ctrl
Cut to Clipboard	Ctrl+X
Duplicate Shape	Ctrl+D
Ellipse Tool.	Ctrl+9
Field	Ctrl+F9
Fill Format.	F3
Flip Horizontal.	Ctrl+H
Flip Vertical	Ctrl+J
Font	F11
Freeform (Spline) Tool	Ctrl+5
Glue (toggles on or off)	F9
Group	Ctrl+G

Help, display online Help F1
Line Tool . . . . Ctrl+6
Line Format . . . . Shift+F3
Macros. . . . . Alt+F8
New Blank Drawing . . . . Ctrl+N
Open Drawing . . . . Ctrl+O
Page F5
Page Properties . . . . Shift+F5
Paragraph . . . . Shift+F11
Paste from Clipboard . . . . Ctrl+V
Pencil Tool . . . . Ctrl+4
Pointer. . . . . Ctrl+1
Print . . . . Ctrl+P
Rectangle Tool . . . . Ctrl+8
Redo . . . . Ctrl+Y
Repeat last Command F4
Rotate Left. Ctrl+L
Rotate Right . . . . Ctrl+R
Rotation . . . . Ctrl+Ø
Save Drawing . . . . Ctrl+S
Save As. F12
Save Workspace . . . . Alt+F12
Select All. . . . . Ctrl+A
Select Shape's Text Block F2
Send To Back. . . . . Ctrl+B
Snap & Glue . . . . Alt+F9
Snap (toggles on or off) . . . . Shift+F9
Spelling F7
Spline (Freeform) Tool . . . . Ctrl+5
Tab Settings . . . . Ctrl+F11
Text . . . . Ctrl+2
Text Height field . . . . Ctrl+Shift+H
Text Rotation angle field. . . . . Ctrl+Shift+A
Text Width field . . . . Ctrl+Shift+W
Tile Windows Horizontally. . . . . Shift+F7
Tile Windows Vertically . . . . Ctrl+Shift+F7
Undo last Command . . . . Ctrl+Z
Ungroup . . . . Ctrl+U
Visual Basic Editor. . . . . Alt+F11
Whole Page . . . . Ctrl+W

```
Zoom. . . . . . . . . . . . . . . . . . . . . . . F6
Zoom Window . . . . . . . . . . . . . . . . . Ctrl + Shift
```

Special Text Characters	Keystroke
Beginning single quote	Ctrl + [
Ending single quote	Ctrl +]
Beginning double quote	Ctrl + Shift + [
Ending double quote.	Ctrl + Shift +]
Bullet	Ctrl + Shift + 8
En dash	Ctrl + =
Em dash	Ctrl + Shift + =
Discretionary hyphen	Ctrl + hyphen
Nonbreaking hyphen.	Ctrl + Shift + hyphen
Nonbreaking slash.	Ctrl + Shift + /
Nonbreaking backslash	Ctrl + Shift + \
Section marker.	Ctrl + Shift + 6
Paragraph marker.	Ctrl + Shift + 7
Copyright symbol	Ctrl + Shift + C
Registered trademark	Ctrl + Shift + R

Hyperlink Navigation (1)	Keystroke
Alt + Right Arrow	Forward
Alt + Left Arrow	Back
Ctrl + Page Down	Next Page
Ctrl + Page Up.	Previous Page

(1) After you add navigational links (jumps) to shapes or pages, use these keyboard shortcuts to navigate between Visio and the jump destination's program.

Keystroke	Function
Alt	Activate the menu bar
Alt + F7	Cascade Windows
Alt + F8	Macros
Alt + F9	Snap & Glue
Alt + F11.	Visual Basic Editor
Alt + F12.	Save Workspace
Ctrl.	Copy shapes
Ctrl + = ...	En dash

Ctrl + [	Beginning single quote
Ctrl +]	Ending single quote
Ctrl + 1	Pointer
Ctrl + 2	Text
Ctrl + 3	Connector Tool
Ctrl + 4	Pencil Tool
Ctrl + 5	Freeform (Spline) Tool
Ctrl + 5	Spline (Freeform) Tool
Ctrl + 6	Line Tool
Ctrl + 7	Arc Tool
Ctrl + 8	Rectangle Tool
Ctrl + 9	Ellipse Tool
Ctrl + Ø	Rotation
Ctrl + A	Select All
Ctrl + B	Send To Back
Ctrl + C	Copy to Clipboard
Ctrl + D	Duplicate Shape
Ctrl + F	Bring Shape To Front
Ctrl + G	Group
Ctrl + H	Flip Horizontal
Ctrl + I	Actual Size
Ctrl + J	Flip Vertical
Ctrl + K	Connect Shapes
Ctrl + L	Rotate Left
Ctrl + N	New Blank Drawing
Ctrl + O	Open Drawing
Ctrl + P	Print
Ctrl + R	Rotate Right
Ctrl + S	Save Drawing
Ctrl + U	Ungroup
Ctrl + V	Paste from Clipboard
Ctrl + W	Whole Page
Ctrl + X	Cut to Clipboard
Ctrl + Y	Redo
Ctrl + Z	Undo last Command
Shift	Constrain to vertical or horizontal (when moving)
Shift + F3	Line Format
Shift + F5	Page Properties
Shift + F7	Tile Windows Horizontally
Shift + F9	Snap (toggles on or off)

Shift + F11 . . . . Paragraph

Ctrl + Shift. Zoom Window

Ctrl + Shift + 6 . . . . Section marker

Ctrl + Shift + 7 . . . . Paragraph marker

Ctrl + Shift + 8 . . . . Bullet

Ctrl + Shift + A . . . . Text Rotation angle field

Ctrl + Shift + C . . . . Copyright symbol

Ctrl + Shift + H . . . . Text Height field

Ctrl + Shift + R . . . . Registered trademark

Ctrl + Shift + W . . . . Text Width field

Ctrl + Shift + /. . . . . Nonbreaking slash

Ctrl + Shift + =. . . . . Em dash

Ctrl + Shift + [. . . . . Beginning double quote

Ctrl + Shift + \. . . . . Nonbreaking backslash

Ctrl + Shift +] . . . . Ending double quote

Ctrl + hyphen . . . . Discretionary hyphen

Ctrl + Shift + hyphen . . . . Nonbreaking hyphen

Ctrl + Shift + F7 . . . . Tile Windows Vertically

F1 . . . . Help, display online Help

F2 . . . . Select Shape's Text Block

F3 . . . . Fill Format

F4 . . . . Repeat last Command

F5 . . . . Page

F6 . . . . Zoom

F7 . . . . Spelling

F8 . . . . Align Shapes

F9 . . . . Glue (toggles on or off)

F10. . . . . Activate the menu bar

F11. . . . . Font

F12. . . . . Save As

Ctrl + F9. . . . . Field

Ctrl + F11 . . . . Tab Settings

Toolbar Icons

Developer Toolbar

Design Mode . . . .
Insert Controls . . . .
Run Macro . . . .
Show Shape Sheet . . . .
Visual Basic Editor . . . .

Page Toolbar

Glue . . . .
Gotot Page . . . .
Next Page . . . .
Previous Page . . . .
Snap . . . .

Shape Toolbar

Align Shapes . . . .
Bring to Front . . . .
Connect Shapes . . . .
Corner Rounding . . . .
Custom Properties . . . .
Distribute Shapes . . . .
Fill Color . . . .
Fill Pattern . . . .
Fill Style . . . .
Flip Horizontal . . . .
Flip Vertical . . . .
Group . . . .
Layout Shapes . . . .
Line Color . . . .
Line Ends . . . .
Line Pattern . . . .
Line Style . . . .
Line Weight . . . .
Rotate Left . . . .
Rotate Right . . . .
Send to Back . . . .

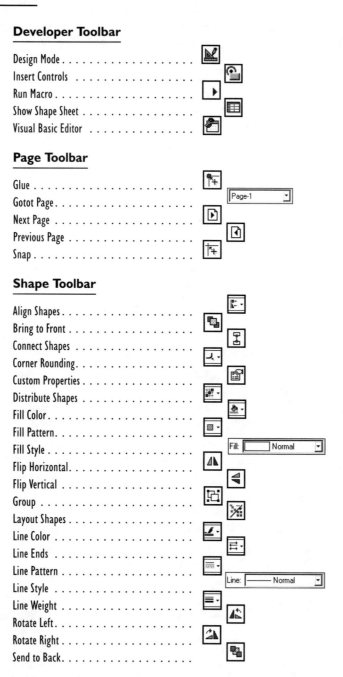

Shadow Color . . . .
Ungroup . . . .

Standard Toolbar

Arc Tool . . . .
Connection Point Tool . . . .
Connector Tool . . . .
Copy . . . .
Crop Tool. . . . .
Cut. . . . .
Ellipse Tool. . . . .
Format Painter. . . . .
Freeform Tool . . . .
Help . . . .
Line Tool . . . .
New . . . .
Open. . . . .
Open stencil . . . .
Paste. . . . .
Pencil Tool . . . .
Pointer Tool . . . .
Print. . . . .
Print Preview . . . .
Rectangle Tool . . . .
Redo . . . .
Rotation Tool. . . . .
Save . . . .
Spelling . . . .
Stamp Tool. . . . .
Text block Tool . . . .
Text Tool . . . .
Undo. . . . .
Zoom In . . . .
Zoom Out . . . .
Zoom Percent . . . .

`58%`

Text Toolbar

Align Bottom . . . .
Align Left . . . .

377

Align Right . . . . ▤ ▤

Align Top . . . . ▥

Bold . . . . **B**

Bullets . . . . ▤

Center . . . . ▤

Decrease Indent . . . . ▦

Decrease Paragraph Spacing . . . . ▦

Decrease Font Size . . . . **A▾**

Font . . . . | ℡ Arial ▾ |

Font Color . . . . **A** ▾

Font Size . . . . | 10pt. ▾ |

Increase Indent . . . . ▦

Increase Paragraph Spacing . . . . ▦

Increase Font Size . . . . **A▴**

Italic . . . . *I*

Justify . . . . ▤

Middle . . . . ▥

Rotate Text . . . . **A↰**

Text Style . . . . | Text: Normal ▾ |

Underline . . . . **U**

View Toolbar

Connections . . . . ▦

Grid . . . . ▦

Gides . . . . ⊤

Layer Properties . . . . ▰

Shape Layer . . . . | {No Layer} ▾ |

Walls Toolbar (*Visio Technical only*)

Align to Match Wall . . . . ▨

Extend to Wall . . . . ▥

Help Wall . . . . **?**

Join Walls . . . . ▤

Move to Wall . . . . ▦

Web Toolbar

Back . . . .

Forward . . . .

Insert Hyperlink . . . .

Search the Web . . . .

Shape Explorer. . . . .

Index